DEVELOPMENTS IN FOOD PRESERVATIVES—1

THE DEVELOPMENT SERIES

Developments in many fields of science and technology occur at such a pace that frequently there is a long delay before information about them becomes available and usually it is inconveniently scattered among several journals.

Developments Series books overcome these disadvantages by bringing together within one cover papers dealing with the latest trends and developments in a specific field of study and publishing them within *six months* of their being written.

Many subjects are covered by the series including food science and technology, polymer science, civil and public health engineering, pressure vessels, composite materials, concrete, building science, petroleum technology, geology, etc.

Information on other titles in the series will gladly be sent on application to the publisher.

DEVELOPMENTS IN FOOD PRESERVATIVES—1

Edited by

R. H. TILBURY

Tate & Lyle Ltd, Group Research & Development, Reading, Berks, UK

APPLIED SCIENCE PUBLISHERS LTD
LONDON

APPLIED SCIENCE PUBLISHERS LTD
RIPPLE ROAD, BARKING, ESSEX, ENGLAND

British Library Cataloguing in Publication Data

Developments in food preservatives.—(The
developments series).
1
1. Food—Preservation
I. Tilbury, R H II. Series
641.4   TP371.2

ISBN 0-85334-918-5

WITH 23 TABLES

© APPLIED SCIENCE PUBLISHERS LTD 1980

Printed in Great Britain by Galliard (Printers) Ltd, Great Yarmouth

PREFACE

The last decade has witnessed significant changes in the attitudes of both private citizens and the legislators towards food additives. They stem from a greater public awareness of the relationships between food and health, coupled with greater official concern over the potential carcinogenicity of chemicals. Food preservatives have not escaped from this critical investigation, with the result that some compounds like diethylpyrocarbonate are no longer permitted, and the maximum levels of others like nitrite and sulphur dioxide have been reduced. This has stimulated the search for newer, safer preservatives, but because of the enormous costs of toxicological safety testing of new chemicals, the main approach has been to find more effective ways of using existing preservatives.

As the title of this book implies, its scope is largely confined to recent developments, i.e. those which have occurred over the last 6–8 years. Prior to this period, scientific advancement in the field of food preservatives had been slow and steady, but recently significant developments have been made in response to the challenges outlined above. This book summarises and reviews these advances. Its timing is appropriate because it is 16 years since a major symposium focused attention on microbial inhibitors in food (4th International Symposium on Food Microbiology, Göteborg, Sweden, 1964).

Deliberately, this book does not aim to be a comprehensive review of the preservation of specific commodities. Rather, it attempts to look at general principles, illustrated by examples of particular foods. The first chapter sets the scene and discusses the methodology of assessing food preservatives. Carole Burke then describes the history of relevant legislation and provides a comprehensive and up-to-date summary of current, world-wide, legislative practice. In Chap. 3, Tim Gray explains the rationale of modern toxicological testing and its interpretation, which is essential to an understanding of the safety in use of food preservatives. He then assesses the current toxicological status of preservatives, which explains some of the legislative trends. No law is effective unless it can be backed up by enforcement procedures, and the latter requires sensitive, specific, accurate and reproducible methods for detection and determination of preservatives in foods. Analytical methods are covered most comprehensively by Sawyer and Crosby in Chap. 4. Their survey demonstrates the major advances that have taken place in both instrumentation and in use of new techniques such as high-pressure liquid chromatography. Professor Sinskey takes a fundamental look at the mode of action of existing and possible future preservatives in Chap. 5. He argues that this knowledge should help us to use preservatives more effectively. Finally, Smith and Pintauro speculate on future trends in the chemical preservation of foods. It seems likely that rather than develop new chemicals, better ways will be found of using approved existing compounds.

R. H. TILBURY

CONTENTS

LIST OF CONTRIBUTORS

CAROLE S. BURKE

    Cadbury Typhoo Ltd, PO Box 171, Franklin House, Bournville, Birmingham B30 2NA, UK.

N. T. CROSBY

    Food Additives and Contaminants Section, Laboratory of the Government Chemist, Cornwall House, Stamford Street, London SE1 9NQ, UK.

T. J. B. GRAY

    The British Industrial Biological Research Association, Woodmansterne Road, Carshalton, Surrey SM5 4DS, UK.

NICHOLAS D. PINTAURO

    Department of Food Science, Rutgers University, New Brunswick, New Jersey 08903, USA.

R. SAWYER

    Food and Nutrition Division, Laboratory of The Government Chemist, Cornwall House, Stamford Street, London SE1 9NQ, UK.

ANTHONY J. SINSKEY

    Department of Nutrition and Food Science, Massachusetts Institute of Technology, Cambridge, Massachusetts 02139, USA.

JAMES L. SMITH

Meat Laboratory, Eastern Regional Research Center, Science and Education Administration Agricultural Research, US Department of Agriculture, Philadelphia, Pennsylvania 19118, USA.

R. H. TILBURY

Tate & Lyle Ltd, Group Research & Development, Philip Lyle Memorial Research Laboratory, University of Reading, PO Box 68, Reading, Berks RG6 2BX, UK.

Chapter 1

INTRODUCTION

R. H. TILBURY

Tate & Lyle Ltd, Reading, UK

'We bless thee for our creation, preservation, and all the blessings of this life.'
    The Book of Common Prayer: A General Thanksgiving
'There is death in the pot.'
    2 Kings iv, 40

Although the Prayer Book quotation above refers to the preservation of human life, rather than to the preservation of food, in practical terms the latter certainly has aided the former. In contrast, the Biblical quotation, 'There is death in the pot.', cited by Accum (1820)[1] in the foreword to his book on chemical adulteration of foods, was intended to illustrate the dangers of using poisonous additives. Then this sentiment was laudable as it paved the way for food legislation which has resulted today in food being safer than at any previous time. However, as pointed out by Jarvis and Burke,[2] the total prohibition of food preservatives is undesirable as it would certainly lead to increased incidence of death caused by growth of food-poisoning organisms. Here then, we have the bones of a dilemma with which this book is largely concerned: How do we maximise the benefits and minimise the risks of preserving food by adding anti-microbial chemicals?

HISTORICAL BACKGROUND

Traditional methods of preserving food, such as drying, salting, syruping, pickling, smoking and fermentation, were established by early civilisations, having evolved empirically from chance discoveries. Detailed accounts of the food eaten in those times were given at a Symposium on the History of

1

Food Science and Technology, presented at the 34th Annual Meeting of the Institute of Food Technologists in 1974.[3] For example, Pariser[4] reports that the ancient Egyptians of the First Dynasty (3000–2778 BC) made cheese, whilst by 2000 BC bread-making, the brewing of beer and wine-making were established. The Ancient Greeks of the 6th and 5th Century BC knew how to preserve meat and fish by salting, and also used vinegar to make fish paste or pickle. Spices and herbs were widely used, together with oil and honey, all no doubt playing a part in preservation of food. Even more detail is available concerning food processing in Ancient Rome, described by Clark and Goldblith.[5] For example, Columella (AD ?–70) wrote accounts of the manufacture and use of vinegar and brine for preservation of herbs and vegetables, meat and fish, and combinations of honey in water and beeswax in water for preserving fruits. Wine must was preserved by pitch or resin, or both, whilst wine itself could be preserved by using in addition compounds such as lye-ash, salt water, marble dust, and a combination of iris, fenugreek and gypsum. Fortunately the modern producer of Chianti does not need to resort to such methods!

Whilst man's economy and way of life remained mainly agricultural and rural, traditional methods of preserving foods met most of his needs. Village communities were almost entirely self-supporting in terms of food supply, and they knew how to preserve seasonal foods and vary the diet so that a balanced nutrition could be maintained throughout the year. In Europe and North America, this balance was changed dramatically by the Industrial Revolution of the late 18th Century. The gradual switch to a largely urban, non-agricultural society brought problems which stimulated the development of large-scale food manufacture and distribution. Initially this encouraged widespread adulteration of foods with chemicals, principally to add bulk or colour to disguise poor-quality materials and unhygienic processing. In turn, this initiated legislation to prevent such adulteration, as described for the UK by Burke in Chap. 2. Similar patterns emerged in the USA, summarised in a series of papers at a symposium entitled '200 Years of Food—A Historical Perspective', presented at the 36th Annual Meeting of the Institute of Food Technologists in 1976.[6] The paper by Middlekauf[7] describes the development of US Food Laws. He cites a horrifying list of adulterants used in the 18th and 19th Centuries: '... there was sulphate of copper in pickles, bottled fruits and preserves; mustard contained lead chromate; sugar confectionery contained boiled potatoes, magnesium and ammonia carbonates, potato flour, pipe clay, gypsum, whiting and ground Derbyshire stone along with chalk, alum and boneashes. Bitter almonds were used to flavour wines, vermilion and red lead to colour the rind of Gloucester cheese. Exhausted tea leaves were bought from hotels, coffee-houses and char-women and mixed in factories with a gum solution. Re-dried, they were coloured with black lead, Prussian Blue, indigo, Dutch Pink, turmeric and poisons like copper carbonate and lead chromate and sold as new tea.'

The legal position regarding food preservatives in the UK was not clearly defined until 1925. Immediately prior to that date, preservatives in common use included alum, boric acid and borax, carbolic acid, creosote, formaldehyde, salicylic acid, thymol, acetic, benzoic and sulphurous acids. Jarvis and Burke[2] commented that 'food manufacturers of that time were perhaps more impressed by the work of Lister on antiseptics than they were by the studies of Pasteur. Very properly most of these 'preservatives' are now totally excluded from foods.' Since 1925 the legislation both in the UK and in most other developed countries has not only specified the foods which are permitted to contain preservatives, but has severely restricted the number of permitted chemicals. Furthermore, it usually specifies the maximum permitted levels of addition and most recently lays down criteria of purity of preservatives. The basic philosophy is that use of a preservative must be justified on the grounds of both safety and technological need.

TYPES OF FOOD PRESERVATIVE

A list of the principal anti-microbial chemicals used in foods is shown in Table 1, although not all are classified as preservatives. For the purposes of this book, the definition of a preservative is taken from the Preservatives in Food Regulations, 1979:[8] 'Preservative' means any substance which is capable of inhibiting, retarding or arresting the growth of micro-organisms or any deterioration of food due to micro-organisms or of masking the evidence of any such deterioration. It excludes

(a) any permitted anti-oxidant,
(b) any permitted artificial sweetener,
(c) any permitted bleaching agent,
(d) any permitted colouring matter,
(e) any permitted emulsifier,
(f) any permitted improving agent,
(g) any permitted miscellaneous additive,
(h) any permitted solvent,
(j) any permitted stabiliser,

TABLE 1
ANTI-MICROBIAL CHEMICALS USED IN FOODS

Chemical	General regulatory status					
	No restriction	Widely permitted	Limited use in many countries	Limited use in few countries	No longer permitted	Not yet approved
Organic acids						
Benzoic acid and its salts		+				
p-Hydroxybenzoic acid esters and their salts (parabens)		+				
Propionic acid and its salts			+			
Acetic acid (vinegar)	+					
Acetates and diacetates			+			
Dehydroacetic acid and its salts				+		
Sorbic acid and its salts		+				
Citric acid and its salts	+					
Lactic acid and its salts	+					
Formic acid and formates					+	
Boric acid and borates					+	
Inorganic acids						
Sulphurous acid and sulphites (SO_2)		+				
Carbonic acid (CO_2)	+					
Salts						
Nitrite (and nitrate)		+				
Sodium chloride	+					

	1	2	3	4	5
Anti-biotics					
Nisin					
Pimaricin				+	
Tetracyclines			+	+	
Formaldehyde		+			
Hexamethylenetetramine (hexamine)			+		
Nitrofurane compounds			+		
Sterilising gases					
Ethylene oxide			+		
Propylene oxide			+		
Hydrochloric acid vapour	+				
Other sterilising agents					
Diethylpyrocarbonate (DEPC)					
Hydrogen peroxide		+			
Ozone	+		+		
Peracetic acid	+				
Post-harvest fungicides					
Biphenyl				*	
2-Hydroxybiphenyl				*	
Thiabendazole				*	
Sugars					+
Alcohol (potable)					+
Polyhydroxyalcohols					
Glycerol					+
Propylene glycol					+

* Residues are permitted on imported fruit.

(k) vinegar,
(l) any soluble carbohydrate sweetening matter,
(m) potable spirits or wines,
(n) herbs, spices, hop extract or flavouring agents when used for
     flavouring purposes,
(o) common salt (sodium chloride), and
(p) any substance added to food by the process of curing known as
     smoking.

In some chapters, however, brief coverage is given to 'traditional'
preservatives excluded from the definition above, where it complements the
text, and mention may be made of some compounds which are no longer
permitted or are yet to be approved.

THE NEED FOR FOOD PRESERVATIVES

The historical changes which initiated the need for more effective means of
preserving food have already received brief mention. The last century has
witnessed enormous technological advances in physical methods of food
preservation by thermal processing (canning, pasteurisation), refrigeration, freezing, concentration, drying and most recently irradiation.
However, this has not precluded the necessity to add anti-microbial
chemicals, for reasons stated by Chichester and Tanner.[9] The worldwide
population explosion has resulted in a crisis of food supply, which demands
that losses be reduced to a minimum. It has been estimated that post-
harvest bio-deterioration of tropical horticultural produce accounts for
losses of about 25 % per annum.[10] Added to this is the fact that countries
with the greatest nutritional needs are also less developed and suffer from
inadequate production, distribution, transportation, storage and preservation facilities; they cannot afford the sophisticated and capital-
intensive technology needed for physical preservation of foods. They need
chemical preservatives which are effective, safe and cheap. Even in
developed countries which can afford sophisticated technology, the
physical methods are not satisfactory for all types of foods. Thermal
processing, especially canning, often has adverse affects on the flavour,
colour and texture of a food product. Freezing may be technologically
attractive but it is relatively expensive and inconvenient. Recent consumer
trends in food must be noted. 'Convenience' foods which demand minimal
preparation and cooking are increasingly being sought. Here the use of

preservatives may help extend the shelf-life of refrigerated or mildly heat-processed foods in aseptic packaging. 'Intermediate moisture foods' (IMF) offer the attractions of mild heat processing, long shelf-life without need for refrigeration, ease of storage and minimal need for cooking.[11] They have been widely used for pet foods for over a decade and for human use have particular attractions for space-flight and the armed forces, but they require the addition of preservatives. Dietary shifts towards more perishable foods such as fruits and vegetables also necessitate increased usage of preservatives, for example post-harvest fungicides and gas packaging. Against this, one must also recognise a trend towards 'health foods', 'natural' unrefined foods and organically grown foods without chemical additives.

Jarvis and Burke[2] summarised the main arguments for and against chemical preservatives. Properly used, there is no doubt that food preservatives can fulfil the three desirable objectives of any preservation process: they maintain keeping quality and stability (prevent spoilage), they maintain nutritional quality, and they control food-poisoning organisms (safety). Although chemical preservatives have been used in the past to cover up the use of spoiled or spoiling foods, inferior raw materials and unhygienic processing, the types and levels presently permitted would not adequately prevent rapid decomposition if growth of spoilage organisms was already advanced.

The most emotive argument against use of preservatives is that of potential toxicity to the consumer. This subject is currently of great concern and is dealt with at length by Gray in Chap. 3. There it is explained how the maximum acceptable daily intake (ADI) of preservatives is usually based on an extrapolation of the no-effect level (NEL) observed in toxicological evaluations; typically a safety factor of $\times 100$ is used. It is worth adding here an important point discussed by Jarvis and Burke,[2] namely that 'widening the classes of foods in which specific preservatives are permitted will result in the potential daily intake (PDI) exceeding significantly the recommended ADI. The ratio PDI/ADI can be used as a relative indication of exposure to particular food additives. In practice, the PDI is a hypothetical value based on all foods in which a permitted preservative can occur; it assumes that (a) the preservative is always present at the maximum permitted level; (b) the specific foods are consumed daily during a lifetime; and (c) that preservatives are not decreased during storage, cooking, etc.[12] For some of the more common preservatives the PDI/ADI ratios have been calculated as: benzoates, 0·82–2·8; sulphur dioxide and sulphites, 2·9–17·0; sorbates, 0·42; nitrates, 0·50; nitrites, 1·2. The higher levels quoted for sulphur dioxide and benzoates relate to the excessive consumption of beverages'.

In the USA, certain preservatives are permitted only in conjunction with good manufacturing practice (GMP), and for this purpose the quantity of preservative is restricted to that which is necessary for the particular situation. Theoretically it can be argued that adoption of GMP renders chemical preservatives unnecessary. In practice, however, technological improvements in processing are not always reliable enough or sufficiently effective in preventing spoilage or risk of food poisoning. Added to this is the fact that the retailer or domestic consumer may subsequently abuse the product during storage and use. Here the addition of preservatives is a desirable safeguard.

In conclusion, it may be generally desirable to prohibit or minimise the addition of preservatives to food, but safety, technological and economic factors often outweigh these considerations.

PROPERTIES OF FOOD PRESERVATIVES

The desirable properties of food preservatives were discussed in a classic paper by Ingram, Buttiaux and Mossel;[13] their conclusions were subsequently endorsed and embodied in a list of ten properties by the International Association of Microbiological Societies (IAMS).[14] Since these principles remain valid today, it is worthwhile reiterating them in detail:

'Antimicrobial here refers only to microorganisms which are for some reason undesirable in the food; and it is emphasized that such additives should not be used to counteract the effects of bad hygiene. Further, the terms "bactericidal" and "bacteriostatic" are here understood to imply action against yeasts and fungi besides bacteria.

1. 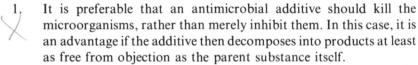 It is preferable that an antimicrobial additive should kill the microorganisms, rather than merely inhibit them. In this case, it is an advantage if the additive then decomposes into products at least as free from objection as the parent substance itself.

2. However, a merely bacteriostatic preservative may be equally effective if it persists until the food is ready for consumption. Similarly, with a food intended for processing, a bacteriostatic additive may be satisfactory if it persists until the food is further processed.

3. On the other hand, additives intended as adjuncts to thermal processes must have an adequate degree of resistance to heat.

4. The range of specificity should correspond with the range of microorganisms able to develop on the food in the likely conditions of use. There is a scarcity of suitable antimicrobial food additives effective in the pH range 5–7, common in foods: new antimicrobial food additives, useful in this pH range, are needed.

5. Any antimicrobial additive should be at least as effective against any food-poisoning species, able to grow in the food under the likely conditions of use, as it is against spoilage microorganisms.

6. Additives intended to supplant thermal processing, wholly or in part, should provide a degree of security against *Clostridium botulinum* similar to that given by the normal thermal process. Recent expert opinion (FAO/WHO/IAEA Joint Committee on the Technical Bases for Legislation on the Wholesomeness of Irradiated Foods, 1964) suggested that this should cause reduction of viable spores by a factor of 10^{12}. Should an additive result in reducing the thermal treatment below this level, other undesirable organisms which might survive must also be adequately controlled. .

7. The additive should not be removed by miscellaneous reactions with the food, nor be inactivated by some specific reversal agent in the food. It should not be destroyed, in concentrations of practical significance, by spoilage microorganisms in numbers likely to occur in good manufacturing practice, nor be inactivated by products of their metabolism.

8. An antimicrobial food additive should not readily stimulate the appearance of resistant strains of microorganisms. Those which do so should be limited to circumstances where the resistant strains are not likely to gain subsequent access to food preserved by the same substance. In both these cases, phenomena of cross-resistance should be taken into account.

9. Certain possible uses of an antimicrobial food additive might be excluded, if the same substance were also used therapeutically or as an additive to animal feeds, and if the former type of use were incompatible with the latter two.

10. There should be a suitable procedure for determining the amount of the additive in foods. Procedures based on microbiological assay have disadvantages which often make chemical and physical methods preferable, if sufficiently sensitive and specific.

TABLE 2

RELATIONSHIP BETWEEN IAMS CRITERIA[a] FOR THE 'IDEAL' PRESERVATIVE AND THE PRESERVATIVES PRESENTLY PERMITTED IN FOODS—JARVIS (1980)[16]

Criterion	Preservative compliance with criterion							
	Benzoate	Parabens	Sorbate	Propionate	Nitrite	Sulphite	Nisin	Pimaricin
Toxicological acceptability	+[e]	+	+	+	?	(+)	(+)	+
Microbiocidal activity	–	–	–	–	–	(+)	(+)	–
Long persistence in food	+	+	+	+	–	–	(+)	+
Chemical reactivity[b]	L	L	M	L	H	H	L	L
High thermal stability[c]	NR	NR	NR	NR	NR	NR	NR	NR
Wide anti-microbial spectrum[d]	YMB	BbYM	YMBb	BM	B	BbYM	B	YM
Active against food-borne pathogens	(+)	–	(+)	(+)	+	(+)	(+)	–
No development of microbial resistance	–	+	–	+	+	–	+	–
Not used therapeutically	+	+	+	+	+	+	+	+
Assay procedure available	+	+	+	+	+	+	+	+

[a] Criteria recommended by IAMS.[14]
[b] Preservatives should not react significantly with food (H = high, M = moderate, L = low reactivity).
[c] Thermal stability important only for preservatives used as thermal adjunct (NR = not relevant).
[d] Ideally the spectrum should be wide (Y = yeasts, M = moulds, B = gram-positive bacteria, b = gram-negative bacteria).
[e] + = comply with criterion; (+) = comply to some degree; – = do not comply; ? = debatable.

Because such matters were outside the competence of the Symposium, the above recommendations do not consider the obvious additional requirements that a food preservative should be free from objection on toxicological, technical and aesthetic grounds. Recommendations on these matters, among other things, are made by the FAO/WHO Codex Alimentarius Commission, for instance'.

To the above list should be added other desirable features of a preservative, essentially non-microbiological, as listed by Kimble:[15]

1. It should be economical and practical.
2. It should be easily soluble.
3. It must be non-toxic.
4. It must exhibit antimicrobial properties over the pH range of each particular food.
5. It must not impart off-flavours when used at levels effective in controlling microbial growth.
6. It must exhibit no carcinogenic properties.

Unfortunately none of the currently permitted preservatives fulfills all these characteristics, as shown in Table 2. A comprehensive account of the properties and uses of each preservative is outside the scope of this book. For more detailed information the articles by Chichester and Tanner[9] and Kimble[15] should be consulted.

CHOOSING A PRESERVATIVE

It has been seen that no single preservative matches up to the ideal requirements. Therefore, in choosing a preservative for a particular application, one must attempt to optimise the desirable characteristics of the available preservatives. In practice, the effectiveness of a preservative depends not only on its inherent properties but on the micro-environment in which it is placed. In a classic paper, Mossel[17] described the factors which influence the growth and survival of micro-organisms in foods. The same ecological approach can be usefully applied to preservatives, which interact with micro-organisms, the food itself, the processing factors and storage conditions. Table 3, taken from Jarvis and Burke[2] neatly summarises this approach. Their paper should be consulted for a detailed discussion of factors affecting preservative efficiency.

TABLE 3
FACTORS AFFECTING PRESERVATIVE ACTIVITY—AFTER JARVIS AND BURKE (1976)[2]

The preservative itself:	anti-microbial spectrum
	solubility
	partition coefficient
	dissociation constant
	reactivity
	relative toxicity
The microflora of the food:	numbers, types and condition of organisms
	resistance to anti-microbial agents
Nature of the food:	water activity
(intrinsic factors)	pH value
	redox potential
	fat content
	reactive components
	nature of ingredients
Processing factors:	thermal processes
	dehydration
	developed preservatives (e.g. acid, smoke, etc.)
Extrinsic factors:	storage conditions (e.g. temperature, relative
	humidity, gaseous atmosphere)
	type of packaging

PRACTICAL EVALUATION OF PRESERVATIVES

The 'Hurdle' Concept

Reliance on 'single' preservation methods for foods, e.g. low temperature, low pH, low water activity, heat, irradiation, or use of one chemical preservative, is not suitable for many foods because of adverse effects on organoleptic or textural properties. In addition, the current trends towards reductions in process levels and reductions in the levels of additives for toxicological reasons, coupled with a consumer demand for more 'natural' foods, necessitates a new approach to food preservation. One approach is to make a fundamental study of the mechanisms of preservation, as discussed by Sinskey in Chap. 5; this embraces studies on the mode of action and location of use of preservatives. Hopefully, this can lead to improvements in the effectiveness of anti-microbial chemicals in foods. An alternative and more empirical approach is to search for favourable interactions between chemical preservatives and physical processes, or

TABLE 4
A_w RANGE AND INHERENT 'HURDLES' OF SOME TRADITIONAL IMF—
LEISTNER AND RODEL (1976)[18]

Products	A_w range	Inherent hurdles
Jams	0·90–0·80	A_w, pH, Eh, F, pres.[a]
Meats	0·90–0·60	A_w, pH, (Eh), t, pres.[a], c.m.[b]
Cake and pastry	0·90–0·60	A_w, (t), F, pres.[a]
Dried fruits	0·75–0·60	A_w, pH, (F), pres.[a]
Frozen foods	0·90–0·60	A_w, t

[a] Preservatives (such as sorbic acid, nitrite).
[b] Competitive microflora (Lactobacillaceae, Streptococcaceae, moulds).

between several preservatives, which will enable reductions in levels to be made without sacrificing the safety or stability of the food. Ideally, synergistic combinations might be found (e.g. benzoic acid and SO_2 in fruit concentrates), but in most cases the effects of each component preservative/process will be additive, each comprising a 'hurdle' to growth of particular types of micro-organisms. This 'hurdle' concept was described by Leistner and Rödel[18] with reference to IMF. Table 4 illustrates this concept in theory whilst Table 5 shows how it can be effectively applied to the preservation of some traditional IMF. As pointed out by these authors, it is possible to inhibit a low number of organisms with fewer and much lower 'hurdles' than are required for the inhibition of a large number of organisms of the same type. This emphasises the need for use of good-quality raw materials and GMP.

It is outside the scope of this book to describe developments in the

TABLE 5
A_w AND INHERENT 'HURDLES' OF SOME NEWLY DEVELOPED IMF—LEISTNER
AND RODEL (1976)[18]

Products	A_w	Inherent hurdles
Sweet and sour pork	0·85	A_w, pH, (Eh), F, pres.[a]
Chicken dish, ready-to-eat	0·85	A_w, F, pres.[a]
Sliced and dried bologna	0·85	A_w, Eh, F, pres.[a]
Hennican	0·85	A_w, (pH), Eh, (F), pres.[a]
Diced carrots	0·77	A_w, (F), pres.[a]

[a] Preservatives (such as sorbic acid, propylene glycol, glycerol, nitrite).

preservation of many specific commodities, but an attempt will be made to discuss the general principles applicable to selection of preservative systems.

Choice of Test System

A major difficulty in the evaluation of preservative interactions is that it is not practicable to measure these in the wide variety of products which may use similar methods of preservation. This is due mainly to the inherent variability of composition observed not only between similar food products but between batches of the same food product, e.g. cured meats. Patel and Gibbs[19] described a method of overcoming this difficulty. The general approach is to use media or model systems for initial assessment of preservative interactions. Results from these are then confirmed in specific foods and finally applied to a wider range of related products. This approach proved effective in reducing the acetic acid levels in pickles and sauces containing salt and sugar;[20] the growth of acetic-acid-tolerant organisms could be prevented by the addition of preservatives or by pasteurisation.[21] In high-pH foods where food-poisoning organisms are likely to grow, assessment of interactions has been more difficult. For

TABLE 6

ADVANTAGES AND DISADVANTAGES OF VARIOUS TEST SYSTEMS FOR THE EVALUATION OF MICROBIAL INHIBITORS—PATEL AND GIBBS (1980)[19]

Parameter	Test system		
	Medium	Model	Food
1. Simplicity of setting up in lab.	+ + +	+	−
2. Reproducibility	+ + +	+ +	−
3. Ease of monitoring samples	+ + +	+ +	(+)
4. Control of other factors	+ + +	(+)	(+)
5. Work with pathogens	+ + +	+	(+)
6. Cheapness	+ + +	+ +	−
7. Similarity to food	(+)	(+)	+ + +

Key:
−   Very poor.
(+)   Poor/variable.
+   Fair.
+ +   Good.
+ + +   Excellent.

example, in work aimed at the reduction of nitrite levels in cured meats, early studies using media showed that *Cl. botulinum* was inhibited by low nitrite levels, particularly after heating (the 'Perigo' effect); this effect could not be observed in real meat systems.[22] Similarly, interactions between A_w and pH for inhibition of food-poisoning organisms in media systems may indicate general trends; actual limiting values differ, however, depending on the nature of the humectant, acid and food type. Final determination of safe limits must be done in specific foods.

The advantages and disadvantages of various test systems for the evaluation of microbial inhibitors are shown in Table 6.

Other Features of Experimental Design

Having chosen the test system to be used for evaluating a preservative system, decisions must be made regarding other important experimental parameters. As a general principle, it is desirable to simulate as closely as possible the actual conditions likely to be experienced during storage and use of the food product in question. In order to be able to guarantee effectiveness of the preservative, these conditions should approach the most adverse likely to be experienced, without being unrealistic.

Micro-organisms

With most test systems, it is unlikely that the presence of the right qualitative and quantitative microflora can be guaranteed, so it is usually necessary to contaminate the food artificially with challenge organisms. In the case of real food systems, however, it is desirable to run some parallel tests with naturally contaminated food.

Where a food product is likely to be stored for an extended time period with multiple samplings by the consumer, e.g. jam, it may be realistic to re-inoculate the product at intervals. This is especially desirable where the persistence of a preservative is in doubt.

Choice of suitable challenge organisms presents some practical problems. For ease of reproducibility and assessment of results, single pure cultures are desirable, but to minimise numbers of samples mixed pure cultures or even mixed, unknown cultures are preferable. Whatever the choice, the organisms should be typical of those likely to occur naturally in the fresh product and should also include known spoilage organisms for the product. Where possible the cultures should have been isolated from similar products. Recent isolates are preferable to cultures which have been maintained in a collection for a long period, as the latter may lose virulence.

Even within a species, considerable variation occurs between strains, e.g.

TABLE 7

VARIATION IN SENSITIVITY OF STRAINS OF *Cl. botulinum* TO PHYSICAL
FACTORS AFFECTING GROWTH—PATEL AND GIBBS (1980)[19]

Growth parameter		Value needed for inhibition of germination and growth of Cl. botulinum	
		Proteolytic strains	Non-proteolytic strains
1. Temperature (°C)	maximum	> 50	> 50
	minimum	< 10	< 3
2. Acidity (pH)	maximum	> 8·5	> 8·5
	minimum	< 4·5	< 5·0
3. Water activity, A_w		< 0·94	< 0·97

proteolytic and non-proteolytic strains of *Cl. botulinum* differ significantly in their sensitivity to physical factors (Table 7).

Clearly, therefore, it is desirable to use several strains of challenge organism. Likewise, for organisms which form spores, it is essential to include spores in the challenge inoculum, especially where heat-processed foods are concerned. Whilst it is obvious that spores are more heat-resistant than vegetative cells of the same strain, it is worth remembering also that chemical preservatives may exert very different effects on spores than on vegetative cells, so that both should be tested.

The physiological state of challenge organisms is important, as exponentially growing, well adapted cells will be more of a spoilage risk than stationary-phase organisms grown in a medium completely different from the test system.

Inoculum size can be a critical factor and one in which decisions are often difficult and subjective. For example, the level of *Cl. botulinum* spores incorporated into model cured meat systems was shown to have a marked effect on the minimum inhibiting concentrations of salt and nitrite.[23] It is sensible to choose realistic inoculum levels based on observation of those found in practice.

Homogeneous mixing of inoculum and substrate is clearly desirable but often difficult to achieve, especially with real food systems.

Duration of Test
The desired shelf-life of a food product may vary from a few days in the case of uncured meat and fish to several years as in the case of IMF. When

developing preservative systems for new products, it is obviously helpful to have results available in as short a time as possible. Conversely, it is known that spoilage may take a long time to develop, because (i) sub-optimal growth conditions retard growth rate; (ii) in thermally processed foods, heat-damaged cells may survive but require time to repair; (iii) in foods of low water activity, the lag phase of organisms may be extended;[24] and (iv) time taken to develop visible spoilage is dependent on initial inoculum size. To resolve this dilemma, it is necessary to devise accelerated ageing tests which predict real shelf-life with sufficient statistical accuracy.

There does not appear to be any reliable scientific basis for developing such a test, but an excellent example of a successful method, evolved on a purely empirical basis, is described by Measures and Cheney.[25] They wished to evaluate preservative systems for IMF petfoods, with a required shelf-life of two years, in the minimum time and with the minimum number

TABLE 8
ACCELERATED AGEING TEST FOR IMF PETFOODS—MEASURES AND CHENEY (1980)[25]

No. of samples	Storage temperature (°C)	Storage relative humidity (%)
3	25	100
3	Ambient	100
3	25	ERH
3	Ambient	ERH (traditional packaging)

of samples. A set of 12 samples, inoculated with 10^3 mould spores/g, are incubated under the conditions of Table 8. It was found that if all 12 samples survive 50 days incubation without appearance of visible mould, there was 95 % probability that the product would survive more than two years under normal conditions. The statistical reliability could be improved by increasing the number of samples tested.

Incubation Temperature
Temperature abuse during storage of foods, both by retailer and consumer, is clearly a risk which is known to occur in practice; therefore shelf-life tests must be done at various temperatures. 'Target' temperatures may be 4 °C for refrigerated products, 15–20 °C for products stored at ambient

temperature in temperate climates and 30 °C for products destined for tropical countries.

It may be desirable to include samples stored under cyclical temperature conditions which simulate day and night temperatures. In such cases, temperature gradients can occur in a product, leading to localised changes in the micro-environment. Significant shifts in A_w may then result from moisture migration, evaporation, etc.

Replication and Numbers of Subsamples
Difficulties in ensuring homogeneous distribution of challenge organisms in the test system, coupled with the possibility of localised differences in the micro-environment mentioned above, require that several subsamples should be examined at each sampling time. Alternatively, replicate samples must be tested.

Resuscitation of Damaged Cells
When determining viable counts of micro-organisms in systems that have been exposed to a stress, the method used should permit resuscitation of sub-lethally damaged cells in order to avoid false-negative results. The principles were outlined by Mossel.[26]

Assessment and Application of Results
Experiments such as those described above to assess preservative interactions in model systems usually generate a mass of data. The challenge is to evaluate these data and present them in a form which is readily comprehensible, establishes the quantitive nature and statistical significance of the interactions, and which enables them to be used to predict the effectiveness of applying such preservative systems in real foods of a diverse nature. Recently a number of advances have been made in solving these problems and will be described briefly here.

Sinskey (Chap. 5) mentions the need to draw stability maps for particular products which depict isosurface responses to predict shelf-life. One approach is to present results in the form of three-dimensional block diagrams. This has been used successfully by Roberts, Jarvis and Rhodes[23] and Jarvis, Rhodes and Patel[27] in their work on the inhibition of *Cl. botulinum* in pasteurised cured meats by curing salts and other additives. Figure 1 illustrates this technique.

Two-dimensional stability maps were drawn by Baird-Parker and Kooiman[28] for products such as soft drinks, jams and preserves. Figure 2 illustrates this technique.

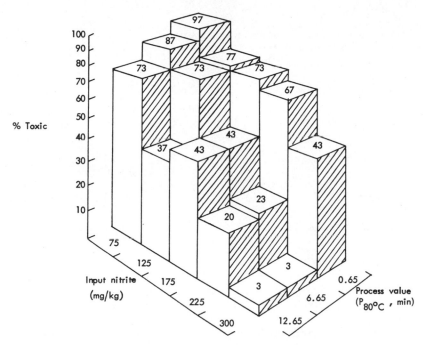

FIG. 1.   The effect of sodium nitrite (mg/kg) and thermal process value ($P_{80\,°C}$, min) on the occurrence of botulinal toxin in meat slurries stored for 183 days at 25 °C. (Data on all salt levels combined.)[27]

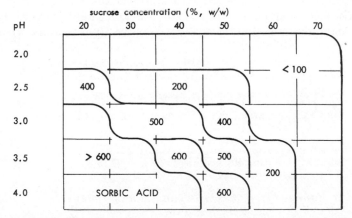

FIG. 2.   Minimum levels (mg/kg) of sorbic acid required to assure an ambient shelf-life of over 3 months, in a model system (malt extract broth) inoculated with xerophilic yeasts.[28]

TABLE 9

RISK FACTORS (R) FOR CHANGES IN CURING AGENT LEVELS AND OTHER
PARAMETERS ASSUMING A 6-MONTH SHELF-LIFE OF PASTEURISED HAM—
JARVIS et al. (1979)[27]

Salt (as % salt/water)	2	3·5	4·5	5·5
R	4	1	0·5	0·1
Nitrite (mg/kg, input)	50	100	200	300
R	10	3	1	0·5
Polyphosphate (% w/w)	0	0·3	0·3	
R	1	1*	0·5–0·2*	
pH value	5·7	6·0	6·4	
R	0·3	1	4	
Process ($P_{80^\circ C}$, min.)	0·7	6·7	12·7	
R	3	1	0·9	
Storage temp. (°C)	15	20	25	
R	1	5	7	

* Dependent on salt, pH and type of phosphate.

Another useful concept developed by Jarvis et al.[27] is that of 'risk factors'. Using data obtained from extensive studies on the safety and stability of pasteurised cured meats, it is possible to indicate some order of magnitude for inhibitory effects of the parameters tested. The 'risk factors' (R) shown in Table 9 are based on a notional value of 1 (unity) for a pasteurised ham containing 3·5 % salt-on-water, 200 mg/kg nitrite, pH 6·0, $P_{80^\circ C}$ 6·7 min and stored at 15°C for 6 months. It is assumed that the

TABLE 10

EXAMPLES OF RISK FACTORS (R) FOR PASTEURISED HAM—JARVIS et al. (1979)[27]

(A) *Average present-day practice*
  Salt (3·5%) + nitrite (200) + P_{80}(6·7) + storage (15°) + pH (6·0)
  $R =$ 1   ×   1   ×   1   ×   1   ×   1   = 1

(B) *Reduce salt and nitrite; increase temperature and pH*
  Salt (2%) + nitrite (50) + storage (20°) + pH (6·4)
  $R =$ 4   ×   10   ×   5   ×   4   = 800

(C) *Reduce nitrite and process; increase salt and temperature*
  Salt (4·5%) + nitrite (100) + storage (20°) + P_{80}(0·7)
  $R =$ 0·5   ×   3   ×   5   ×   3 $\simeq$ 23

(D) *As (C) + polyphosphate*
  $R = 23$ ×   0·2 $\simeq$ 5

parameters interact and that the R factors can be multiplied to give a combined risk factor (Table 10). To illustrate this concept, a reduction in salt to 2 %, and nitrite to 50 mg/kg in a meat with a post-process pH value of 6·4, storage at 15 °C would result in a 160-fold increase in risk. Storage of such a product at 20 °C would result in an 800-fold risk increase (Table 10, B). If the salt level in meat at pH 6·0 were increased to 4·5 %, nitrite reduced to 100 mg/kg and the process to be marginal ($P_{80\,°C}$ 0·7), then the risk factors associated with storage at 20 °C would be increased some 23-fold (Table 10, C). However, by incorporating an inhibitory phosphate (Table 10, D), the risk factors could be limited to a 5-fold increase. Although this approach cannot be precise, it may indicate the possible order of magnitude of potential hazards with which manufacturers could be faced in certain circumstances. In a situation where consumers require less salty (more bland) foods, legislators press for lower nitrite levels, and manufacturers face problems of high pH meats and the need to maximise yield by controlling losses, etc., some guidance of risk potential is essential. However, it is not yet possible to predict the 'safe' limiting levels for all types of cured meats.

Finally, computers can be used to evaluate preservative interactions and derive models which can predict shelf-life of products. An excellent example of this approach is described by Measures and Cheney[25] with respect to the preservation of IMF petfoods. Their objective was to study the shelf-stability of semi-moist petfoods in A_w range 0·75–0·92 with varying levels of preservatives and with varying pH. The primary test, described in the previous section (Table 8) used mould spores as challenge organisms; subsequently, samples found stable against moulds were checked against potential yeast and bacterial spoilage organisms. Over 1200 products were manufactured and tested; the analyses of the products, and their shelf-life under the various test conditions, were loaded onto a computer and manipulated by a system sub-package of statistical routines. It became clear that asking the computer to correlate all the data at once was expensive and liable to produce less than best results. The task was made simpler by breaking down the data into subsets, each with some of the variables held constant. Simple correlations were then made within the subsets, and the results were correlated as sets of equations. At the end of this manipulation, it was possible to define an equation which described the stability of a product. A product is stable if its A_w is less than the following expression of preservative concentrations:

$$A_w < k_1 + p \times \text{(propylene glycol)} + \frac{q}{\exp(k_2 - \text{pH})} \times \text{(potassium sorbate)}$$

where (i) k_1 is a constant; being the value of A_w required for a stable product when no preservative is present; (ii) p is a coefficient describing the efficacy of propylene glycol in this system, and it is constant with pH; (iii) q is a coefficient describing the efficacy of potassium sorbate in this system, and it is modified with pH by the expression $1/\exp(k_2 - pH)$, where k_2 is related to the pK of potassium sorbate; (iv) use of this equation is subject to the following limits:

$$A_w < 0.87 \qquad pH\ 3.5\text{--}7.5$$

0.1% sorbate or its equivalent must be present in semi-moist products ($A_w\ 0.75\text{--}0.87$).

In no case was synergy found between any of the preservatives or between a preservative and A_w or pH (other than described by dissociation of acids). Some examples of stable products derived from the above equation are shown in Table 11.

TABLE 11
PRESERVATION PARAMETERS NEEDED FOR STABLE SHELF-LIFE
OF IMF PETFOODS—MEASURES AND CHENEY (1980)[25]

A_w	pH	Propylene glycol (%)	Potassium sorbate (%)
0.847	6.0	2.0	0.20
0.798	5.8	1.5	0.15
0.825	6.5	2.3	0.12
0.787	4.5	1.0	0.15
0.797	4.2	0.0	0.25

Computer modelling appears to offer great promise as an aid to optimisation of preservative systems in new food and beverage systems. Hopefully the next decade will see this tool used to the maximum benefit of mankind.

ACKNOWLEDGEMENTS

The author is grateful to the Directors of Tate and Lyle Limited for permission to publish this chapter. Especially I wish to thank Basil Jarvis for his wise counsel over the last decade, and in particular for his helpful advice and discussions in planning this book. I have drawn largely upon his work and that of his staff at the Food Research Association for much of my contribution, and had time enabled him, he would have been a major contributor himself. Finally, may I thank the authors of each chapter for the invaluable part they have played in this book.

REFERENCES

1. ACCUM, F. (1820) *A treatise on adulterations of food and culinary poisons*, Longman, Hurst, Rees, Orme and Brown, London.
2. JARVIS, B. and BURKE, C. S. (1976) 'Practical and legislative aspects of the use of food preservatives'. In: *Inhibition and inactivation of vegetative microbes*, F. A. Skinner and W. B. Hugo (Eds.), Academic Press Ltd, London, p. 345.
3. Symposium on the History of Food Science and Technology (1975) *Fd Technol.*, **29**(1), 22–53.
4. PARISER, E. R. (1975) 'Foods in Ancient Egypt and Classical Greece', *Fd Technol.*, **29**(1), 23–7.
5. CLARK, J. A. and GOLDBLITH, S. A. (1975) 'Processing of foods in Ancient Rome', *Fd Technol.*, **29**(1), 30–2.
6. Symposium: 200 years of food—A historical perspective (1976) *Fd Technol.*, **30**(6), 30–61.
7. MIDDLEKAUF, A. D. (1976) '200 years of U.S. food laws—A Gordian knot', *Fd Technol.*, **30**(6), 48–54.
8. Preservatives in Food Regulations (1979) S.I. No. 752, H.M.S.O., London.
9. CHICHESTER, D. F. and TANNER, F. W. (1972) 'Antimicrobial food additives', In: *Handbook of Food Additives*, 2nd edn, T. E. Furia (Ed.), The Chemical Rubber Co., Cleveland, Ohio, Chap. 3, p. 115.
10. COURSEY, D. G. (1972) 'Biodeteriorative losses in tropical horticultural produce', In: *Biodeterioration of Materials*, Vol. 2, A. H. Walters and E. II. Hueck-van der Plas, (Eds.), Applied Science Publishers Ltd.; London, pp. 464–71.
11. DAVIES, R., BIRCH, G. G. and PARKER, K. J. (Eds.) (1976) *Intermediate Moisture Foods*, Applied Science Publishers, London.
12. Codex Alimentarius Commission (1975) *Potential Intake of Food Additives*, Paper CX/FA 75/5 prepared by Codex Secretariat, WHO/FAO, Rome.
13. INGRAM, M., BUTTIAUX, R. and MOSSEL, D. A. A. (1964) 'General microbiological considerations in the choice of antimicrobial food preservatives', In: *Microbial Inhibitors in Food*, N. Molin (Ed.), Almqvist and Wiksell, Stockholm, Göteborg, Uppsala, pp. 381–92.
14. MOLIN, N. (Ed.) (1964) *Microbial Inhibitors in Food*, Almqvist and Wiksell, Stockholm, Göteborg, Uppsala, pp. 393–4.
15. KIMBLE, C. E. (1977) 'Chemical food preservatives'. In: *Disinfection, Sterilisation and Preservation*, 2nd edn, S. S. Block (Ed.), Lea and Febiger, Philadelphia, pp. 834–58.
16. JARVIS, B. (1980) 'Chemical food preservation: An overview'. In: Proc. Soc. Chem. Ind. Joint Symp., Food Group and Microbiology, Fermentation and Enzyme Technology Group, *Recent Developments in Antimicrobial Food Preservatives*, London, 12th December 1979. *J. Sci. Fd Agric.*, in press.

17. MOSSEL, D. A. A. (1971) 'Physiological and metabolic attributes of microbial groups associated with foods', *J. Appl. Bacteriol.*, **34**(1), 95–118.
18. LEISTNER, L. and RÖDEL, W. (1976) 'The stability of intermediate moisture foods with respect to microorganisms'. In: *Intermediate Moisture Foods*, R. Davies, G. G. Birch and K. J. Parker (Eds.), Applied Science Publishers, London, pp. 120–37.
19. PATEL, M. and GIBBS, P. A. (1980) 'Laboratory assessment of preservative interactions'. In: Proc. Soc. Chem. Ind. Joint Symp., Food Group and Microbiology, Fermentation and Enzyme Technology Group, *Recent Developments in Antimicrobial Food Preservatives*, London, 12 December 1979. *J. Sci. Fd Agric.*, in press.
20. MUYS, G. T. (1971) 'Microbial safety in emulsions', *Process Biochem.*, **6**(6), 25–8.
21. BINSTEAD, R., DEVEY, J. D. and DAKIN, J. C. (1971) In: *Pickle and Sauce Making*, Food Trade Press, London.
22. ASHWORTH, J. and SPENCER, R. (1972) 'The Perigo effect in pork', *J. Fd Technol.*, **7**, 111.
23. ROBERTS, T. A., JARVIS, B. and RHODES, A. C. (1976) 'Inhibition of *Clostridium botulinum* by curing salts in pasteurised pork slurry', *J. Fd Technol.*, **11**, 25.
24. HORNER, K. J. and ANAGNOSTOPOULOS, G. D. (1973) 'Combined effects of water activity, pH and temperature on the growth and spoilage potential of fungi', *J. Appl. Bacteriol.*, **36**, 427–36.
25. MEASURES, J. C. and CHENEY, P. A. (1980) 'I.M.F. petfood preservation'. In: Proc. Soc. Chem. Ind. Joint Symp., Food Group and Microbiology, Fermentation and Enzyme Technology Group, *Recent Developments in Antimicrobial Food Preservatives*, London, 12 December 1979. *J. Sci. Fd Agric.*, in press.
26. MOSSEL, D. A. A. (1975) 'Water and Microorganisms in foods—A synthesis'. In: *Water Relations of Foods*, R. B. Duckworth (Ed.), Academic Press, London, pp. 347–61.
27. JARVIS, B., RHODES, A. C. and PATEL, M. (1979) 'Microbiological safety of pasteurised cured meats: Inhibition of *Clostridium botulinum* by curing salts and other additives'. In: *Food Microbiology and Technology*, B. Jarvis, J. H. B. Christian and H. D. Michener (Eds.), Medicina Viva, Parma, Italy, p. 253.
28. BAIRD-PARKER, A. C. and KOOIMAN, W. J. (1980) In: *Microorganisms in Foods 3, Microbial Ecology of Foods, Vol. II, Food Commodities*, The International Commission on Microbiological Specifications for Foods, Academic Press, London.

Chapter 2

INTERNATIONAL LEGISLATION

CAROLE S. BURKE

Cadbury Typhoo Ltd, Birmingham, UK

HISTORICAL BACKGROUND TO THE NEED FOR LEGISLATION

The evolution of legislation on food follows the evolution of an industrial society. In a mostly rural community with food being consumed in the vicinity of its production, the needs for preservation could be met by traditional methods such as cheesemaking, salting, curing, drying, smoking, fermentation and preserving with sugar. These methods are insufficient for an industrialised country because the movement of the population to the towns necessitates the transportation of food from the farms to the cities. This brings with it the problems of the preservation of perishable fresh foods. Until the advent of refrigeration and pasteurisation, the only means of protecting fresh foods from spoilage was the addition of chemical preservatives. Those available in the nineteenth century included boric, acetic, carbolic, benzoic, salicylic and sulphurous acids and formaldehyde.

Unfortunately, the addition of preservatives and other chemical additives to foods was not only for the justifiable reason of food preservation. The adulteration of food was becoming a source of concern. F. C. Accum's *Treatise on Adulteration of Food and Culinary Poisons* published in 1820[1] gave proof of this activity but it was very difficult to achieve legislative reform because of public ignorance, opposition from the influential trade sector and inadequate law enforcement.

25

The law cannot be enforced without suitable analytical methods for the detection of adulteration which will stand up in the courts. Dr A. H. Hassell published such methods of detecting adulteration in the reports of *The Lancet* Analytical Sanitary Commission in 1851–1854. These reports received sufficient publicity to arouse public agitation for legislation to prohibit adulteration. A Parliamentary Select Committee Inquiry into adulteration was held in 1855–1856 and as a result of their deliberations a Bill was introduced in 1857. However, it was hastily withdrawn in the face of strong opposition from interested parties. Anti-adulteration pressure groups were formed, in particular the Social Science Association in Birmingham. Then an incident occurred which gave these pressure groups the ammunition they needed. In 1858, in Bradford, 200 people were poisoned, 17 of them fatally, from eating adulterated lozenges. This was followed by the first general Act of Parliament to prevent adulteration— the Adulteration of Food and Drink Act 1860.[2]

Unfortunately, this Act contained no effective enforcement measures and consequently it did not achieve its objective. Public pressure continued to mount and there was even a journal published called the *Anti-Adulteration Review*.

The ineffectual 1860 Act was replaced by the Sale of Food and Drugs Act 1875, 'An Act to make better provision for the sale of food and drugs in a pure state'. However, it was not until the amendment of 1879 that this Act became enforceable. This was mostly due to a change in attitude of the trade who could now see that it was to their advantage to prevent unfair practices as this encouraged fair competition.

A further act was passed in 1899 followed by the Food and Drugs (Adulteration) Act 1928 which repealed all previous acts. The 1928 Act was mainly concerned with dairy products, as milk was still a major source of food poisoning and tuberculosis and subject to adulteration. This 1928 Act included the prohibition to 'mix, colour, stain or powder any article of food with any ingredient or material so as to render the article injurious to health'.[3]

The Food and Drugs Act 1955 as amended by the Medicines and Pharmacy Act 1968 (to remove the drugs from food and drugs) is that currently in force. It has recently been under review and the Ministry of Agriculture, Fisheries and Food announced by press notice in January 1980 that it has been decided that no significant changes to the Act are necessary.

The basic requirements of this Act are fundamental to all food law. These are that (*a*) food must not be rendered injurious to health, and (*b*) food must be of the nature, substance and quality demanded by the consumer.

LEGISLATION SPECIFICALLY ON PRESERVATIVES

The United Kingdom

The first legislation in this country specifically on preservatives came after the Ministry of Health produced a report in 1924 which led to the passing of the Public Health (Preservatives, etc., in Food) Regulations S.R. & O. 1925 No. 775 (last amended in 1927 S.R. & O. No. 577). These regulations listed 17 foods which could contain preservatives and prohibited any other foodstuff from containing any added preservative. The declaration of the presence of preservative was also required on the foodstuff. The permitted preservatives were benzoic acid and sulphur dioxide.

A definition of preservative was included:

any substance capable of inhibiting, retarding or arresting the process of fermentation, acidification or other decomposition of food or of masking any of the evidence of putrefaction but does not include common salt (sodium chloride), saltpetre (sodium or potassium nitrate), sugars, lactic acid, acetic acid or vinegar, glycerine, alcohol or potable spirits, herbs, hop extracts, spices and essential oils used for flavouring purposes, or any substance added to food by the process of curing known as smoking.

This means that all of the substances such as salt which are excluded from the definition of preservatives (i.e. defined out) are not subject to these regulations and may be used freely in any food. 'Freely' of course means within the restrictions of the Food and Drugs Act that the food should not be injurious to health and must be of the nature, substance and quality demanded by the consumer.

This was a great step forward as the legislation protected the consumer from the addition of injurious substances such as formaldehyde and protected the honest trader from unfair competition. At least it would be an achievement if the regulations were complied with, but the effectiveness of these regulations was not immediate. This is to be expected as it takes time for new laws to become familiar to the manufacturer and to the enforcer. Today, means of communication regarding new laws are far more effective. Manufacturers belong to trade associations who are actively involved both in the drawing up of new legislation and in the distribution of information concerning existing legislation.

There have been several amendments to the first regulations on preservatives since 1925 but the fundamental principles have remained the same. In 1940 sodium nitrite was permitted at 200 mg/kg in bacon, ham and pickled meat (sodium nitrate was still defined out and could therefore be

TABLE 1(a)

FOODS PERMITTED TO CONTAIN PRESERVATIVES (PRESERVATIVES IN FOOD REGULATIONS 1979)

Food	Maximum levels (mg/kg)						
	Sulphur dioxide		Benzoic acid		Parabens[f]		Sorbic acid
Beer	70	and	70	or	70		
Beetroot, cooked & prepacked			250	or	250		
Candied peel or cut & drained (syruped) peel	100						
Cauliflower, canned	100						
Cheese							1 000
Chicory & coffee essence			450	or	450		
Cider	200					or	200
Coconut, desiccated	50						
Colouring matter except E150 caramel, if in the form of a solution of a permitted colouring matter			2 000	or	2 000	or	1 000
The permitted colouring matter E150 caramel	1 000						
Desserts, fruit-based milk and cream	100			or			300
Dessert sauces, fruit-based with a total soluble-solids content of less than 75%	100	or	250	or	250	or	1 000
The permitted miscellaneous additive, dimethylpolysiloxane	1 000	or	2 000	or	2 000	or	1 000

Item						
Enzymes:						
Papain, solid	30 000					
Papain, aqueous solutions	5 000	or				1 000
Aqueous solutions of enzyme preparations not otherwise specified, including immobilised enzyme preparations in aqueous media	500	or	3 000	or	3 000	
Figs, dried	2 000	or	3 000		500	
Finings when sold by retail:						
Wine finings	12 500					
Beer finings	50 000					
Flavourings	350	or	800	or	800	
Flavouring syrups	350	or	800	or	800	
Flour for biscuits[a]	200					
Flour confectionery[b]						1 000
Foam headings, liquid	5 000	or	10000	or	10000	
Freeze drinks	70	or	160	or	160	
Fruit-based pie fillings	350	or	800	or	800	
Fruit, crystallised or glacé or drained	100	or	1000	or	1000	
Fruit, dried other than prunes or figs	2 000	or	1 000	or	300	
Grapes	15			or	450	
Fruit juices:[c]			1000	or	1 000	
Fruit juice and concentrated fruit juice obtained in each case from apples, grapefruit, oranges or pineapples[c]	50					
Fruit juice and concentrated fruit juice obtained in each case from grapes[c]	50 (or after 17/11/1979, 10)					
Fruit juice and concentrated fruit juice obtained in each case from lemons or limes[c]	350					

TABLE 1(a)—contd.

Food	Sulphur dioxide		Benzoic acid		Parabens[f]		Sorbic acid
			Maximum levels (mg/kg)				
Any other fruit juice or concentrated fruit juice	350	or	800	or	800		
Fruit (other than fresh fruit) or fruit pulp, including tomato pulp, paste or purée	350	or	800	or	800		1 000
Fruit spread	100	and					
Garlic, powdered	2 000						
Gelatin	1 000						
Ginger, dry root	150						
Glucose drinks containing not less than 23·5 lb of glucose syrup per 10 gallons of the drink	350	or	800	or	800		
Grape juice products (unfermented, intended for sacramental use)	70	and	2 000	or	2 000		
Grape juice, concentrate, intended for home wine making and labelled as such	2 000						
Hamburgers or similar products	450						
Herring, marinated, whose pH does not exceed 4·5[d]			1 000	or	1 000		
Hops, dried, sold by retail	2 000						
Horseradish, fresh grated & horseradish sauce	200	or	250	or	250		
Jam (other than diabetic jam)	100						
Jam, diabetic	100	and	500	or	500	or	1 000

Product	Permitted level(s)
Low-fat products consisting of an emulsion principally of water in oil	
Mallow, chocolate covered	2 000 / 1 000 (of mallow and chocolate together)
Mackerel, marinated, whose pH does not exceed 4·5[d]	50
Mushrooms, frozen	100 or 1 000 or 1 000
Nut pastes, sweetened	
Olives, pickled	100 or 250 or 250 or 1 000 or 500
Peas, garden, canned, containing no added colouring matter	250
Pectin, liquid	200
Perry	100 or 250 or 250 or 200 or 1 000
Pickles, other than pickled olives	100
Plants (including flowers and seeds), crystallised, glacé or drained	100
Potatoes, raw, peeled	50
Preparations of permitted artificial sweetener & water only	2 000
Prunes	750 and 250 or 1 000
Salad cream (including mayonnaise) and salad dressing	100 or 250 or 250 or 1 000
Sambal oelek	850 and 1 000
Sauces, other than horseradish sauce	100 or 250 or 250 or 1 000
Sausages or sausage meat	450
Soft drinks for consumption after dilution not otherwise specified in this Schedule	350 or 800 or 800 or 1 500
Soft drinks for consumption without dilution not otherwise specified in this Schedule	70 or 160 or 160 or 300

TABLE 1(a)—contd.

Food	Maximum levels (mg/kg)			
	Sulphur dioxide	Benzoic acid	Parabens[f]	Sorbic acid
Starches, including modified starches	100			
Sugars:[e]				
Specified sugar products:				
1. Dextrose anhydrous, dextrose monohydrate, extra white sugar, semi-white sugar, white sugar	15			
2. Invert sugar solution, invert sugar syrup, sugar solution	15			
3. Icing dextrose, icing sugar	20			
4. Soft sugar, white soft sugar	40			
5. Glucose syrup sold otherwise than by retail for use in the manufacture of sugar confectionery and other foods	400			
6. Glucose syrup not specified in item 5	20			
7. Dried glucose syrup sold otherwise than by retail for use in the manufacture of sugar confectionery and other foods	150			
8. Dried glucose syrup not specified in item 7	20			
Hydrolysed starches (other than specified sugar products)	400			
Other sugars except lactose	70			

Tea extract, liquid	450						
Vegetables, dehydrated:							
Brussels sprouts	2 500						
Cabbage	2 500						
Potato	550						
Others	2 000						
Vinegar:							
Cider or wine vinegar	200						
Other	70						
Wine (including alcoholic cordials) other than wine in respect of which the maximum permitted sulphur dioxide content is prescribed by any Community Regulation	450		450	and	200		
Wine in respect of which the maximum permitted sulphur dioxide content is prescribed by any Community Regulation	As prescribed in milligrams per litre by the Community Regulation						
Yogurt, fruit	60	or	120	and	120	or	200 / 300

[a] Bread and Flour Regulations 1959 SI No. 2106.
[b] May contain propionic acid (see Table 1(b)) or sorbic acid.
[c] Fruit Juices and Fruit Nectars Regulations 1977 SI No. 927.
[d] See also Table 1(b).
[e] Specified Sugar Products Regulations 1976 SI No. 509.
[f] See footnote to Table 1(b).

added). Further amendments in 1962 increased the number of foods which may contain permitted preservatives and the number of permitted preservatives was increased to include sorbic acid and *para*-hydroxybenzoic acid esters in addition to *ortho*-phenylphenol, propionic acid and tetracyclines (these latter were later prohibited), and restricted the use of

TABLE 1(*b*)

FOODS PERMITTED TO CONTAIN PRESERVATIVES (PRESERVATIVES IN FOODS REGULATIONS 1979)

Food	Preservative[a]	Maximum level (mg/kg)
Bread	Propionic acid	3 000 of flour
Cheese other than Cheddar, Cheshire, Grana-padano or Provolone type cheeses or soft cheese[c]	Sodium nitrate or Sodium nitrite	100 10
Provolone cheese	Hexamine	25 (as formaldehyde)
Christmas pudding	Propionic acid	1 000
Flour confectionery[d]	Propionic acid	1 000
Fruit, fresh:		
Bananas	Thiabendazole	3
Citrus fruit	Biphenyl or	70
	2-hydroxybiphenyl or	12
	Thiabendazole	10
Herring, marinated, whose pH exceeds 4·5[d]	Hexamine and Benzoic acid or Parabens[b]	50 1 000 1 000
Mackerel, marinated, whose pH exceeds 4·5[d]	Hexamine and Benzoic acid or Parabens[b]	50 1 000 1 000
Meat, cured (including bacon or ham)[e]	Sodium nitrate and sodium nitrite	500 200

[a] When two or more preservatives are permitted in a food, both may be used together only if, 'when the quantity of each preservative present in that food is expressed as a percentage of the maximum quantity of that preservative appropriate to that food in accordance with the Schedule, the sum of those percentages does not exceed one hundred' (see p. 46).
[b] Parabens = methyl-, ethyl- and propyl-4-hydroxybenzoate and their sodium salts.
[c] Proposals to reduce the levels of nitrate and nitrite (see p. 48).
[d] See also Table 1(*a*).

saltpetre to bacon, ham, pickled meat and some cheeses (no limit was laid down).

The regulations were revised in 1974 to increase further the use of preservatives and to introduce purity criteria for the preservatives themselves.

The preservatives regulations currently in force are the Preservatives in Foods Regulations 1979 SI No. 752. The definition of preservative has been changed slightly to make it more precise but it is fundamentally the same: 'preservative' means any substance which is capable of inhibiting, retarding or arresting the growth of micro-organisms or any deterioration of food due to micro-organisms or of masking the evidence of any such deterioration but does not include (a) any permitted anti-oxidant, (b) any permitted artificial sweetener, (c) any permitted bleaching agent, (d) any permitted colouring matter, (e) any permitted emulsifier, (f) any permitted improving agent, (g) any permitted miscellaneous additive, (h) any permitted solvent, (j) any permitted stabiliser, (k) vinegar, (l) any soluble carbohydrate sweetening matter, (m) potable spirits or wines, (n) herbs, spices, hop extract or flavouring agents when used for flavouring purposes, (o) common salt (sodium chloride), or (p) any substance added to food by the process of curing known as smoking.

The foods permitted to contain preservatives under these regulations are given in Tables 1(a) and 1(b).

Other Countries

The history of food legislation in other countries is similar to that in the United Kingdom because the basic requirements for the prevention of disease, the preservation of food and the protection of the consumer are the same.

Table 2 shows a comparison between preservatives permitted in 1933[4] and 1979 in selected countries. This table shows the differences between countries as well as the differences between time. This difference between countries will be elaborated further on.

As for the difference between 1933 and 1979, it is generally similar in all countries. Preservatives such as boric acid and salicylic acid have been replaced by sulphur dioxide and sorbates. The number of foods which are permitted to contain preservatives has increased, and maximum levels have been introduced.

The reasons for these changes are:

(a)   the continued industrialisation and urbanisation increasing the need for improved keeping qualities of foods;

TABLE 2

A COMPARISON OF PRESERVATIVES IN 1933[4] AND 1979 IN SELECTED COUNTRIES IN A FEW FOODS—MAXIMUM PERMITTED LEVELS (mg/kg)

Food		Belgium	Denmark	France	West Germany	Italy	Netherlands	UK	Norway	Sweden	Canada	USA
Baked goods	1933										B, SO_2, Bo, Sa	B, SO_2
	1979	1000 S	1000 S			2000 S		1000 S	2000 S	2000 S	1000 S	GMP S
Cheese, including processed	1933			Bo[a], Sa[a]		2000 Bo					B, SO_2, Bo, Sa	B
	1979	1000 S	1000 S			1000 S		1000 S	1000 S	2000 S	3000 S, 2000 P	2000 S, 3000 P
Jam	1933	SO_2 tr			B[c], F[c]		250 Sa[b], 250 B[b]	40 SO_2	SO_2 tr	B	B, SO_2, Bo, Sa[b]	B, SO_2
	1979	1000 S, 40 SO_2	500 B, 1000 S, 100 SO_2	20 SO_2	B[c], S[c], 50 SO_2	80 SO_2	250 B, 250 S, 50 SO_2	100 SO_2	1500 B, 1000 S, 50 SO_2	1000 B, 2000 S, 50 SO_2	1000 B, 1000 S, 500 SO_2	1000 B, GMP S, GMP SO_2
Marzipan	1933			1500 B							B, SO_2, Ba, Sa	B, SO_2
	1979		1500 S		1500 S	1000 S		1000 S	S	1000 S	1000 S	GMP S
Pickles	1933							250 B			B, SO_2, Ba, Sa	B, SO_2
	1979	300 B, 300 S, 50 SO_2	1000 S, 1000 B		2000 B, 1500 S, 20 SO_2	50 SO_2		100 SO_2, 250 B	1500 B, 50 SO_2, 1000 S	2000 B, 200 SO_2, 2000 S	1000 B, 1000 S, 500 SO_2	1000 B, GMP S, GMP SO_2
Soft drinks containing fruit juice	1933				500 B	200 SO_2[d]	20 SO_2, 250 B, 250 Sa	350 SO_2, 600 B	SO_2	B	B, SO_2, Ba, Sa	B, SO_2
	1979	100 B, 100 S	200 B, 500 S, 25 SO_2		1000 B, 1000 S, 10 SO_2	160 B, 20 SO_2	250 B, 250 S, 100 SO_2	350 SO_2, 800 B	1500 B, 1000 S, 35 SO_2	1000 B, 1000 S, 50 SO_2	1000 B, 1000 S, 100 SO_2	1000 B, GMP S, GMP SO_2

P = propionates.
S = sorbates.
Sa = salicylic acid.
SO_2 = sulphur dioxide.
B = benzoic acid.
Bo = borates.
F = formaldehyde.
tr = trace.
GMP = good manufacturing practice.

[a] In rennet.
[b] Second-grade only.
[c] On surface only.
[d] In syrups.

(b) the development of the food industry from small manufacturers with no means of proper control over the addition of preservatives to larger manufacturers with adequate resources to ensure proper control—more preservatives can be permitted if it can be safely assumed that they will not be misused;

(c) improvements in analytical methods have made it possible to specify and enforce maximum levels;

(d) the consumer demand for a variety of foods to be available out of season and for foods to have longer shelf lives; and

(e) the introduction of new foods for which traditional methods of preservation are inadequate, e.g. low-sugar jams and liquid artificial sweeteners.

THE MECHANISM OF LEGISLATION

How Legislation is Drawn up in the United Kingdom

For convenience, the United Kingdom is referred to as if there is only one set of legislation. In fact, as far as food laws are concerned, there are three sets of laws: those for England and Wales, those for Scotland, and those for Northern Ireland. The procedure for preservatives is usually as follows. The Ministry of Agriculture, Fisheries and Food (MAFF) jointly with the Department of Health and Social Security (DHSS) announce by press notice that the preservatives regulations are to be reviewed or that proposals for amendment of the existing regulations are to be circulated. In the case of a review, comments are invited for presentation to the Food Additives and Contaminants Committee (FACC) which consists of invited representatives from the enforcement authorities, from industry, from the health authorities, from the academic world and from consumer groups. This committee gathers comments from all interested parties and accepts written and oral evidence. It also consults any other sources it deems necessary. The report of the committee's review is then published. The Ministers are not committed to accept all or any part of this report and its recommendations are sometimes rejected or only accepted in part. It is a common fallacy that an FACC report must be complied with as if it were law. However, in the absence of any law it is a useful guide, but the FACC recommendations do not override existing legislation.

After the report has been published, the MAFF and DHSS collect comments on the report from all interested parties. If the need for new legislation or amendments to existing legislation has been established, then proposals for regulations are circulated by the relevant Ministry to interested parties such as trade associations, enforcement associations, research associations and consumer groups. A specified period of time is allowed for comments on the proposals. If the proposals are very contentious, further amended proposals may be then circulated, but usually the next step is for the regulations themselves to be laid before Parliament. A further press notice announces the publication of the new regulations which normally include a transitional period before they come into force to allow manufacturers to adjust their procedures to comply with the new requirements.

In cases of emergency, where it may suddenly come to the notice of the authorities that a particular practice is dangerous, they may dispense with the above procedure and go directly to the stage of issuing regulations which come into effect immediately. There is usually sufficient prior warning, however.

The procedure for passing a new Act or amending an existing Act is different from regulations in that it involves the Parliamentary procedure for Acts of Parliament including the publication of a Bill, etc., rather than the simple 'laying before Parliament' which is all that is required for regulations. The principles of consultation of interested parties is the same.

Food regulations are subject to regular review as a requirement of the Food and Drugs Act. The use of a preservative must be justified on the grounds of safety and technological need. If its use in a particular food or in general can no longer be justified, then that use or the preservative itself will be removed from the permitted list. Similarly, if a new use of an existing preservative or a new preservative can satisfy the criteria of safety and need, then it will be added to the permitted list.

The Enforcement of Legislation in the United Kingdom
The enforcement of food legislation in the United Kingdom is effective because of the close cooperation between the food manufacturers, the Government and the enforcement authorities. The legislation is drawn up by the Government after thorough consultation with all interested parties. Therefore, when the regulations are published, their content is already known to and accepted by the manufacturers (although they may not necessarily always be welcomed). The Acts and regulations are enforced by the local authorities' Trading Standards or Consumer Protection Departments who have the power to enter the premises of the manufacturer or seller, to carry out inspection and to remove samples. It is the local authority which brings a prosecution in the Courts against offenders.

A law cannot be enforced if there is no means of proving that an offence has been committed. The local authority takes regular samples, usually from shops, of foods and sends them to the Public Analyst who is not only qualified as an analytical chemist but is also an expert on food law. If the Public Analyst's report is unfavourable, the local authority contacts the manufacturer and asks for an explanation. If there is no satisfactory explanation, events could lead to a prosecution after formal samples have been taken and examined. Offences against the Food and Drugs Act and regulations thereunder are criminal offences, not civil, and are liable to imprisonment in some cases.

The position in other countries is similar to that in the United Kingdom although the detail can vary considerably.

The 'Carry-over' Principle

Legislation on preservatives throughout the world has the common factor of a restricted list of foods which may contain a restricted number of preservatives at specified maximum levels of use. It would be illogical to allow a preservative in a particular food but prohibit that preservative when that food is used as an ingredient of another food which may not contain the preservative. It is therefore usual to allow a 'carry-over' of preservatives in permitted foods into other foods which may not contain preservatives when the permitted food is used as an ingredient of that other food. For example, jam may contain sulphur dioxide but cakes may not. However, a cake with a jam filling may contain sulphur dioxide by virtue of its being 'carried-over' in the jam. The amount of sulphur dioxide must be in proportion to the amount of jam. Thus, as jam may contain 100 mg/kg of sulphur dioxide and as the jam-filled cake contains, for example, 25 % jam, then the finished cake may contain 25 mg/kg of sulphur dioxide.

INTERNATIONAL PRESERVATIVES LEGISLATION

The following Tables 3–8 show the permitted preservatives in selected countries. The countries have been grouped politically and geographically:

(a) EEC countries: Belgium, Denmark, France, West Germany, Italy, Netherlands, United Kingdom;

(b) Scandinavia: Norway, Sweden;

(c) North America: Canada, USA.

These countries were selected as they are the most influential with regard to

---

TABLE 3

PRESERVATIVES PERMITTED FOR FOOD USE. GENERAL INDICATIONS OF THE AVERAGE LEVELS (mg/kg) PERMITTED IN SELECTED COUNTRIES (SEE TABLES 4–8).

Food	Benzoates	Parabens	Sorbates	Propionates	Sulphur dioxide (range)[a]
CEREAL PRODUCTS					
Baked goods			1 000	2 000	
Bread			1 000	3 000	
DAIRY PRODUCTS					
Cheese, including processed			1 000	2 000	
Liquid egg, whole or yolk	10 000		10 000		
FISH PRODUCTS					
Fish semi-preserves	2 000	1 000	2 000		
FRUIT AND VEGETABLE PRODUCTS					
Candied fruit			1 000		50–500
Dried fruit					100–2 500
Dried vegetables					250–2 500
Fruit juice	1 000	1 000	1 000		10–3 000
Fruit pulp	1 000	1 000	2 000		+
Jam	1 000	1 000	1 000		20–500
Mustard, prepared	1 000	1 000	1 000		10–500
Pickles	1 000	1 000	1 000		20–500
Sauces, spiced	1 000	1 000	1 000		10–500
Soft drinks containing fruit juice	1 000	1 000	1 000		10–350
Soft drinks, flavoured, usually carbonated	1 000	1 000	1 000		15–100
OLEAGINOUS FOODS					
Margarine	2 000		1 000		
Marzipan			1 000		
Mayonnaise	1 000	1 000	1 000		

[a] Levels vary so greatly that there is no general level.
+ No maximum permitted level.

TABLE 4

MAXIMUM PERMITTED LEVELS (mg/kg) FOR SULPHUR DIOXIDE (SULPHITES AND THEIR SALTS) IN FOODS IN SELECTED COUNTRIES

Food	Belgium	Denmark	France	West Germany	Italy	Netherlands	UK	Norway	Sweden	Canada	USA
Candied fruit	2 000		GMP	100	100		100	50[o]	200	500	GMP
Dried fruit		1 000	2 000	+[e]	100		2 000	2 000	1 000[i]	2 500	GMP
Dried vegetables	250[f]	1 000	1 000	500[f]			2 000	2 000	1 000	2 500	GMP
(often does not include potatoes)			100	300[g]	1 000[h]		50[m]	10	3 000[i]	500	
Fruit juice	10			+	+	100	+[n]	200	50	+	GMP
Fruit pulp[a]	+			50	80	75	100	50	50	500	GMP
Jam	40	100[l]	20	10		50[b]			200		GMP
Mustard, prepared	250		500	20					200		GMP
Pickles	50[k]			10	50			50	200	500	GMP
Sauces, spiced				10			100				GMP
Soft drinks containing fruit juice	25	25				100	350[c]	35	50	100	GMP
Soft drinks, flavoured, usually carbonated					20	75	70[d]	35	50	100	GMP

[a] Includes pulp for further processing.
[b] Second-grade jam, 75 mg/kg.
[c] For consumption after dilution.
[d] For consumption without dilution.
[e] Apricots, pears, peaches: 2 000 mg/kg. Pineapples, apples, quinces: 1 500 mg/kg. Grapes, excluding currants: 1 000 mg/kg.
[f] Certain specified vegetables only.
[g] Citrus only and not for direct consumption; 10 mg/l in retail grape and orange juice.
[h] For further processing; 50 mg/kg for products for direct consumption.
[i] Apricots, peaches, pears: 2 000 mg/kg. Apples: 1 500 mg/kg.
[j] Lemon juice only.
[k] Onions only.
[l] Only from use of dried fruit.
[m] 350 in some fruit juices.
[n] 350 in pulp for direct consumption.
[o] Cocktail cherries only.
+ No maximum permitted level.
GMP In accordance with good manufacturing practice.

TABLE 5

MAXIMUM PERMITTED LEVELS (mg/kg) FOR BENZOATES (BENZOIC ACID AND SODIUM BENZOATE) IN FOODS IN SELECTED COUNTRIES

Food	Belgium	Denmark	France	West Germany	Italy	Netherlands	UK	Norway	Sweden	Canada	USA
Fish semi-preserves	1 000	1 000		2 500	1 500	5 000		5 000	2 000	1 000	1 000
Fruit juice		200[g]		1 000[a]		250	800[i]		1 000	1 000	
Fruit pulp		+				250	800		1 000	+	1 000
Jam		500		+[a]		250[c]		1 500	1 000	1 000	1 000
Liquid egg, whole or yolk	10 000	500		10 000		12 500			10 000		
Margarine						2 000			2 000	1 000	2 000
Mayonnaise	1 000	1 000		2 500		1 000		3 000	2 000	1 000	1 000
Mustard, prepared	250	1 000		1 500		250		1 500	2 000		1 000
Pickles	300[h]	1 000		2 000			250	1 500	2 000	1 000	1 000
Sauces, spiced	1 000	2 000		2 500		1 000	250	1 500	2 000	1 000	1 000
Soft drinks containing fruit juice	100	200				250	800[d]	1 500	1 000	1 000	1 000
Soft drinks, flavoured, usually carbonated	100	200	160[f]	1 000[b]	160	75	160[e]	1 500	1 000	1 000	GMP

[a] Surface treatment only
[b] Of the base.
[c] Only in second-grade jams.
[d] Soft drinks for consumption after dilution.
[e] Soft drinks for consumption without dilution.
[f] Carbon dioxide minimum 4·7 g/l and pH 3 maximum.
[g] Not for direct consumption.
[h] Certain vegetables only.
[i] Mostly only for further processing into other foods.
GMP In accordance with good manufacturing practice.
+ No maximum permitted level.

TABLE 6

MAXIMUM PERMITTED LEVELS (mg/kg) FOR SORBATES (SORBIC ACID AND ITS SALTS) IN FOODS IN SELECTED COUNTRIES

Food	Belgium	Denmark	France	West Germany	Italy	Netherlands	UK	Norway	Sweden	Canada	USA
Baked goods	1 000	1 000			2 000		1 000	2 000	2 000	1 000	GMP
Bread[a]	1 000			2 000		1 000		2 000	2 000	1 000	GMP
Candied fruit		1 000			1 000			1 000[a]	2 000	1 000	GMP
Cheese, including processed cheese	1 000	1 000			1 000		1 000	1 000[b]	2 000	3 000	2 000
Fish semi-preserves	3 000	1 000		2 000	1 000	5 000		2 000	2 000		GMP
Fruit juice		500[e]		2 000[c]		250					
Fruit pulp	+	+		2 000		250			1 000	1 000	GMP
Jam	1 000	1 000		+[b]		250[c]		1 000	2 000	+	GMP
Liquid egg, whole or yolk	10 000	5 000		10 000		12 500				1 000	
Margarine	1 000			1 200	1 000	2 000	1 000		2 000	1 000	1 000
Marzipan		1 500		1 500	1 000		1 000	+	1 000	1 000	GMP
Mayonnaise	1 000	1 000		2 500		1 000		1 000	2 000		GMP
Mustard, prepared	250	1 000		1 000		250		1 000	2 000	1 000	GMP
Pickles	300[f]	1 000		1 500			1 000	1 000	2 000	1 000	GMP
Sauces, spiced	1 000	1 000		2 500		1 000	1 000	1 000	2 000	1 000	GMP
Soft drinks containing fruit juice	100	500		1 000[d]		250	300	2 000	1 000	1 000	GMP
Soft drinks, flavoured, usually carbonated	100	500				75	300	1 000	1 000	1 000	GMP

[a] Certain types only, mostly sliced wrapped bread
[b] Surface treatment only.
[c] Second-grade jam only.
[d] In the base.
[e] Not in fruit juice for direct consumption.
[f] Certain vegetables only.
[g] Cocktail cherries only.
GMP In accordance with good manufacturing practice.
+ No maximum permitted level.

TABLE 7

MAXIMUM PERMITTED LEVELS (mg/kg) FOR PARABENS (*para*-HYDROXYBENZOIC ACID ESTERS AND THEIR SALTS) IN FOODS IN SELECTED COUNTRIES

Food	Belgium	Denmark	France	West Germany	Italy	Netherlands	UK	Norway	Sweden	Canada	USA
Fish semi-preserves	1 000	300		1 000	1 000		800[f]	500	500	1 000	
Fruit juice		200[c]					800		1 000	1 000	
Fruit pulp		+							1 000	+	
Jam		300						900	1 000 *	1 000	GMP
Mayonnaise		300		1 200	1 000[a]				1 000	1 000	
Mustard, prepared		300		1 500			250	900	1 000	1 000	
Pickles		300					250	900	1 000	1 000	
Sauces, spiced	1 000[b]	300		1 500			800[d]	900	1 000	1 000	
Soft drinks containing fruit juice		200					160[e]	900	1 000	1 000	
Soft drinks, flavoured, usually carbonated		200							1 000	1 000	GMP

[a] Of the fat content.
[b] pH more than 5.
[c] Not for direct consumption.
[d] For consumption after dilution.
[e] For consumption without dilution.
[f] For further processing only.
GMP In accordance with good manufacturing practice,
+ No maximum permitted level.

TABLE 8
MAXIMUM PERMITTED LEVELS (mg/kg) FOR VARIOUS PRESERVATIVES IN FOODS IN SELECTED COUNTRIES

Preservative	Food	Belgium	Denmark	France	West Germany	Italy	Netherlands	UK	Norway	Sweden	Canada	USA
PROPIONATES	Baked goods		3 000			2 000		1 000	1 000	3 000	2 000	GMP
	Bread[b]	3 000	3 000	5 000	3 000	2 000	3 000	3 000	5 000	3 000	2 000	3 200
	Cheese, including processed					+[d]					2 000	3 000
NITRATES	Cured meat		500[g]	+	500	250	2 000	500[j]				
NITRITES	Cured meat	200[e]	75[h]	150	+[e]	150	500	200[j]	60[e,g]	200[e]	200	200[i]
PIMARICIN	Cheese rind	500		+		+	+	+		+		
NISIN	Processed cheese	2.5	200	+		+				+		
ETHYLENE OXIDE	Spices	50					50	+[c]				50
HEXAMINE	Fish and/or caviar[d]	1 000	500	1 000				50		500	GMP	
ACETATES	Ca²⁺ or Na⁺ Bread		+	4 300	+	4 000	4 000	+	+	+[f]	3 000[g]	4 000

[a] Certain specified products only.
[b] Certain types only, mostly sliced wrapped.
[c] As organicide or fungicide only.
[d] Treatment of rind only.
[e] Must be mixed with sodium chloride before use.
[f] Sodium diacetate no limit; calcium acetate 6 000 mg/kg.
[g] Certain products only.
[h] 25 mg/kg in fully preserved products.
[i] 120 mg/kg in bacon + 550 mg/kg ascorbate: see text.
[j] Recommendations for lower limits: see text.
GMP In accordance with good manufacturing practice.
+ No maximum permitted level.

legislation on food and as they are representative of the legislation in the rest of the world.

Every effort has been made to ensure that the information given here is as accurate as possible but it is essential that the following is taken into consideration when using these tables:

(*a*) no details of actual legislation have been given and these tables must *not* be used for production purposes;

(*b*) legislation changes frequently—these tables represent the position as at January 1980 and could be out of date within months;

(*c*) the maximum levels are usually for the food as sold unless otherwise stated—the levels are given for the acid (or for sulphur dioxide), and if the salts are used, higher levels may be employed equivalent to the acid (or to sulphur dioxide);

(*d*) where two or more preservatives are permitted in one specified food, the maximum levels for each which may be used are reduced accordingly—for example, if a food may contain 1000 mg/kg benzoic acid or 200 mg/kg sulphur dioxide, and 750 mg/kg of benzoic acid (i.e. 75 %) is used, then 50 mg/kg of sulphur dioxide (i.e. 25 %) may also be used, but if 1000 mg/kg of benzoic acid is used, then no sulphur dioxide may be used;

(*e*) the tables are not a comprehensive list of all the foods which may contain the preservatives mentioned or of all the permitted preservatives—the most common foods and preservatives have been selected to give a good representation of international legislation on preservatives for food use;

(*f*) an empty space in the table means that the preservative is not permitted in that food in that country.

It can be seen from the tables that the permitted levels of preservatives varies widely between countries. This is because the needs for preservatives differ in each country due to different eating patterns, differences in the national food industries and differences in the attitudes of national governments. For example, in France for many years the addition of any additives to bread was prohibited. As French bread is traditionally consumed within a few hours of baking, there was no need for preservatives. With the advent of sliced wrapped bread which is expected to have a shelf life of at least two days, the laws in France were changed to permit propionic acid in sliced wrapped bread only. The traditional French bread may still not contain preservatives.

Another difference between countries is the number of foods which may contain preservatives. In Belgium, France, West Germany, Italy, the Netherlands and the United Kingdom, the restricted list of foods which may contain permitted preservatives is precise. In Denmark, Norway, Sweden, Canada and the USA there are two lists of foods. There are those foods which are specifically mentioned as permitted to contain preservatives and there is a second class of foods, often called 'non-standardised foods', which does not contain a specific list of foods but consists of all those foods not mentioned elsewhere.

Generally speaking, the non-standardised foods are those which are *not* any of the following:

infant and baby foods, dietetic foods, milk and milk products, meat and meat products, fish and fish products, poultry and poultry products, oils and fats and their products, fruits and vegetables and their products, grain and cereal products including bread and flour confectionery, salt, vinegar, tea, coffee, cocoa, chocolate, soft drinks, and alcoholic beverages.

The preservatives permitted by Denmark, Norway, Sweden, Canada and the USA in these non-standardised foods varies but the general picture is:

1000 mg/kg maximum benzoic acid,
1000 mg/kg maximum sorbic acid,
1000 mg/kg maximum propionic acid,
1000 mg/kg maximum *para*-hydroxybenzoic acid esters.

It is noticeable that in the United States many of the foods may contain permitted preservatives at levels which are not specified except as GMP (in accordance with good manufacturing practice). This must surely create difficulties for the enforcement authorities and for the manufacturers to come to an agreement as to what level of use constitutes good manufacturing practice.

THE FUTURE FOR PRESERVATIVES LEGISLATION

Over the past decade or so there have been no dramatic changes in preservatives legislation but recent events may indicate the prognosis for preservatives. Preservatives by their very nature are harmful—at least to micro-organisms. With advances in toxicology and analytical methods for the detection of minute traces of carcinogens, it is inevitable that preservatives should come under the scrutiny of the toxicologist and that some may not survive this investigation. With preservatives it is essential to balance the risks from microbial food-poisoning against the risks from carcinogens.

An excellent example of this is the proposed reduction in permitted levels of nitrates and nitrites in bacon rind and other cured meats. It has been shown that nitrosamines are carcinogenic and that it is possible for them to be formed in meat cured with nitrites. On the other hand, it is known for certain that there is a positive risk of fatal *botulinum* food-poisoning if nitrite is not present in bacon. Various national governments are therefore trying to decide what action to take to protect the consumer best.

In the USA a proposal to reduce the maximum level of nitrite from 200 mg/kg to 156 mg/kg was introduced in 1974 and this was replaced by a further proposal in 1975 to permit 120 mg/kg maximum of nitrite together with 550 mg/kg of ascorbate in bacon. In 1978 regulations were published[5] prohibiting nitrate in bacon and permitting maximum 120 mg/kg sodium nitrite (148 mg/kg potassium nitrite) to be added to bacon with the compulsory addition of 550 mg/kg sodium ascorbate or erythorbate. A sampling scheme was also set up to gather analytical information on the presence of nitrosamines in bacon.

Bacon found to contain confirmable levels of nitrosamine (10 μg/kg) is deemed to be adulterated. Bacon manufacturers are required to carry out research to develop methods of curing bacon with as little as 40 mg/kg of nitrite. A proposal was published simultaneously[6] which would permit 40 mg/kg nitrite plus 0·26% potassium sorbate and 550 mg/kg sodium ascorbate or erythorbate to be followed by 12 months' studies by manufacturers under actual processing plant conditions.

At about the same time there was also published a proposal to permit bacon made without nitrate or nitrite.[7] This product could still be called bacon if it had a similar flavour and consistency of traditionally prepared bacon and if the declaration 'No nitrate, no nitrite. Not preserved, must be refrigerated below 40 °F' was made on the product. This latter declaration was not required if the product had been preserved by thermal processing, pickling, fermentation, had a pH of 4·6 or less, was dried to a water activity of 0·92 or less, etc. This proposal caused much concern over the risk of *botulinum* food-poisoning and it was subsequently revised to the extent that such products must be labelled 'uncured' and 'keep frozen'.[8]

In the United Kingdom the MAFF issued a press notice in August 1978 stating that the question of a reduction in the permitted levels of nitrate and nitrite was under consideration by the FACC. In December 1978 a further press notice announced the publication of the FACC *Report on the Review of Nitrate and Nitrites in Cured Meats and Cheese*. The FACC recommended in its report that the maximum level of nitrates and nitrites should be reduced to the levels shown in the table.

Food	Maximum levels (mg/kg)	
	Nitrate + nitrite	of which Nitrite
Uncooked ham, uncooked bacon	500	200
Sterile packs of cured meats	150	50
Cured meat sausages	400	50
All other cured meats including cooked bacon and ham	250	150

The MAFF are not obliged to accept these recommendations and it does not follow that they will become law. It does seem likely that some action will be taken in due course, but it will not be hasty action.

The use of sulphur dioxide is also likely to be further restricted in the future. Maximum permitted levels have been consistently reduced with each annual revision of the list of permitted additives in the Scandinavian countries with the ultimate aim of removing it all together. The permitted level of sulphur dioxide has been reduced in some sugars and fruit juices and will be reduced in jams and marmalades in the United Kingdom, all as a result of EEC directives.

Not all preservatives are under increased restriction. Sorbic acid is being considered in a number of countries as an alternative to those preservatives now considered to be less safe, such as sulphur dioxide and nitrite. In 1979 the United Kingdom added 18 new permitted uses for sorbic acid to the permitted list.

Although national governments and international organisations such as the European Communities and the Codex Alimentarius Commission of the WHO/FAO are under constant pressure from certain groups to restrict the number and quantity of additives in foods further, they also have to balance the possible harmful effects of the absence of preservatives. Preservatives must not be used as a substitute for sound hygienic practices by the manufacturer but the consumer is accustomed to expect extended shelf lives for many products such as soft drinks for consumption after dilution. In such cases, the preservatives are needed to protect the food from spoilage after the consumer has opened the container.

It is because of the risk/benefit balance which must be achieved with regard to preservatives that it does not seem likely that the number of permitted preservatives in general will change very much. However, as new foods come onto the market, they may well need to contain preservatives and so the uses of the preservatives may be extended.

SUMMARY

The purpose of legislation on preservatives in food is to protect public health and to prevent fraud. The first food laws attempting to achieve this aim were passed in the mid-nineteenth century. The current preservatives regulations in the United Kingdom are explained in detail and the principles can be applied to the laws of other countries, even though they differ in detail.

Most countries have a list of permitted preservatives and no other preservatives may be used in food. The following substances are usually permitted: benzoic acid, propionic acid, sorbic acid, sulphur dioxide and their salts. Most countries also restrict the use of permitted preservatives by specifying that only certain named foods may contain them and laying down maximum levels.

When permitting a preservative in a food, governments must be sure that the use of such a preservative is really necessary and it is not used to hide poor hygiene or poor quality raw materials. Governments must also be sure that the protection of health gained from the use of the preservative is not outweighed by the dangers of health presented by the preservative itself.

Some preservatives, such as sulphur dioxide and nitrite, have recently been further restricted because of doubts about their safety. The use of sorbic acid is increasing in some countries. As new foods come onto the market the uses for preservatives could be extended. The future for preservatives legislation does not indicate much change other than this.

ACKNOWLEDGEMENT

I should like to thank the Leatherhead Food RA for the use of their very comprehensive collection of international food legislation.

REFERENCES

1. ACCUM, F. C. (1820) Treatise on Adulteration of Food and Culinary Poisons, Longmans, London.

2. PAULUS, I. (1974) *The Search for Pure Food. A Sociology of Legislation in Britain.* Martin Robertson Law in Society, Barleyman Press, Bristol.
3. ROBINSON, R. A. (1931) *Bell's Sale of Food and Drugs*, 8th edn, Butterworth & Co. Ltd and Shaw & Sons Ltd, London.
4. HINTON, C. L. (1933) *A Summary of Food Laws and Regulations*, The Newman Press Ltd, London.
5. ANON. (1978) *Fed. Register*, **43**(95), 20992.
6. ANON. (1978) *Fed. Register*, **43**(95), 21007.
7. ANON. (1978) *Fed. Register*, **43**(83), 18195.
8. ANON. (1978) *Fd Chem. News*, **20**(37), 4.

National Legislation Consulted for Tables 1–8:
Belgium: 'Arrêté ministériel fixant la liste des additifs autorisés dans les denrées alimentaires 30/8/1976', *Moniteur belge* (1976)¹ **146**(224), 14721.

Canada: *Office Consolidation of the Food and Drugs Act and of the Food and Drugs Regulations with amendments to May 18, 1978*, Department of National Health and Welfare Canada, Minister of Supply and Services, Canada.

Denmark: *Fortegnelse over godkendte tilsaetningsstoffer til levnedsmidler (positivlisten)*, Statens Levnedsmiddelinstitut, Oct. 1977.

France: *Journal officiel de la République française* (1975) **107**(232), 10310. Dehove, R. A. (1978) *La Réglementation des Produits Alimentaires et Autres. Qualité et répression des Fraudes*, 9th edn, Commerce Éditions, Paris.

West Germany: Zipfel, W. (1978) *Lebensmittelrecht*, 35th supplement, C. H. Beck, Munich.

Italy: 'Disciplina degli additivi chimici consentiti nella preparazione e conservazione delle vivande', *Decreto Ministeriale*, 31/3/1965, as last amended 28/7/78.

The Netherlands: *Besluit van 22/12/1967* (Stb. 691) *houdende vastelling van het Conserveermiddelenbesluit (Warenwet), zoals gewijzigd bij besluit van 9/6/1978* (Stb. 375). de Weever, P. L. and Bleijs, H. T. M. (1978) *Warenwet*, Ministerie van Sociale Zaken en Volksgezondheid, Vermande Zonon, Uitgevers.

Norway: *Fortegnelse over de for 1979 godkjente tilsetningsstoffer*, Sosialdepartmentet Helsedirektoratet.

Sweden: *Statens livsmedelsverks kungörelse on livsmedelstillsatser*, SLV FS 1977:6, SLV FS 1978:12.

United Kingdom: O'Keefe, J. A. *Bell and O' Keefe's Sale of Food and Drugs*, 14th edn, 1968 & Issue No. 7, 1977, Butterworth & Co. Ltd and Shaw & Sons Ltd, London. *The Preservatives in Foods Regulations*, 1979, SI No. 752, HMSO, London.

USA: *Code of Federal Regulations*, Title 21, *Food and Drugs*, Department of Health, Education and Welfare, Washington, USA.

Chapter 3

TOXICOLOGY

T. J. B. GRAY

British Industrial Biological Research Association, Carshalton, UK

INTRODUCTION

Chemical preservation of food has its origins in antiquity. Foods preserved by traditional methods, such as curing, pickling and fermentation, are still widely prepared but the more recent introduction of specific chemical preservatives, selected for their anti-microbial activity, has greatly increased the scope of food preservation. Traditionally processed foods, in general, find ready acceptance by regulatory authorities. The same cannot be said for foods processed by addition of chemical preservatives for, despite the very obvious benefits accruing from the use of preservatives, it is essential to ensure that the chemicals used do not, in themselves, constitute a hazard to human health. This is the object of toxicological testing. In this chapter, current approaches to toxicity testing are outlined and the toxicological status of the major preservatives is briefly reviewed.

METHODS IN TOXICOLOGY

The use of food additives in the UK is based on the principle of permitted lists. This means that an additive can be used only after the submission of adequate toxicological data supporting its safety in use to the relevant governmental body. Guidelines indicating the type of data required were published by the Ministry of Agriculture, Fisheries and Food[1] in 1965 and a revised version is currently being prepared by the Department of Health and Social Security. Recommended procedures are also issued jointly by the Food and Agriculture Organisation and the World Health Organisation (FAO/WHO).[2]

These publications constitute recommendations only and not rigid, enforceable sets of standard tests to be performed in every case. This is important, for no single pattern of tests could be universally appropriate with a group of chemicals so structurally and functionally diverse as food additives. The nature and extent of testing necessary to provide a reasonable assurance of safety may differ widely for different chemicals and is a decision calling for experienced toxicological judgement. The range of studies that may be required is outlined in this section.

Preliminary Information

Before initiating toxicity tests in animals, the substance to be tested should be thoroughly defined. The test sample must naturally be representative of the material intended for commercial use and its chemical purity and identity should be known. Careful scrutiny of the chemical structure of a new substance may reveal features in common with substances of well characterised biological activity and this may indicate particular lines of investigation. Both the nature and amounts of any impurities present in the test material should be known since these can significantly influence toxic potential. Chemical properties and physical characteristics such as odour and volatility will determine whether test diets of adequate stability and acceptability to the animals can be formulated, or whether it may be necessary to administer the substance, for example, by stomach tube. As well as characterising the compound itself, it is important to have an estimate of the likely level of human consumption. This will guide both the selection of appropriate dose levels for animal studies and interpretation of the results of these studies.

Acute Toxicity Studies

For a new substance, animal studies will normally begin with an acute toxicity test, that is a study of the effects produced by a single exposure to the test material. For a food additive, the basic procedure involves oral administration of the substance to groups of laboratory animals over a range of dose levels extending from one producing zero to one producing 100% mortality.[3] From the relationship between dose and mortality, the dose level causing 50% mortality, the so-called LD_{50}, can be calculated (LD = lethal dose). The LD_{50} provides a crude basis for comparing the acute toxicity of different materials.

Merely determining an LD_{50}, however, has limited value.[4,5] This is particularly so for food additives, whose acute toxicity is usually very low and which are, in any event, consumed only at low levels but over prolonged periods. An acute toxicity test can yield much more information if the observations made are extended to include the onset, nature and duration of toxic symptoms as well as mortality. These observations, together with gross and microscopic examination of tissues from animals that die and from survivors, may provide some pointers towards the nature and target areas of potential toxicity. Using several species of animal, and both sexes, may show up species and sex differences in sensitivity to the test compound. All of this information will help the planning of longer-term studies.

It is very rare for a potential food additive to be so acutely toxic, in relation to its expected human intake, as to warrant its rejection at this stage. Normally, then, testing will continue with investigations of the metabolism of the compound and longer-term animal studies.

Metabolic Studies
Studies under this heading are designed to establish what happens to the test substance after it is administered to animals and ultimately to man. Information is sought on the extent and rate of absorption of the chemical from the gut, its subsequent distribution within the body and possible accumulation in organs and tissues, the metabolic transformations it may undergo and the mode and rate of its excretion. Clearly the scope of metabolic studies is very wide, but detailed investigations will not always be required. A few of the more immediately relevant aspects are mentioned here and reviews are available.[3,6]

Metabolism may greatly influence the toxicity of a chemical and so, for a new substance, it is important to establish the pattern of metabolites produced. Generally, metabolites are less biologically active and are more readily excreted than the parent compound. If it can be shown that metabolism yields products already characterised as toxicologically innocuous, the need for further testing may be greatly reduced. Conversely, metabolism can give rise to reactive chemical species capable of mediating toxic effects. Where there is evidence of this, or where metabolites known to be associated with particular toxic effects are identified, further studies will be necessary to characterise the circumstances of active metabolite formation. These may lead to rejection of the test substance or at least indicate the need for long-term animal studies.

The metabolism of a chemical may differ both qualitatively and quantitatively in different species. Data on the chemical's metabolism in several species, including man, are therefore highly desirable. Such data are valuable in assessing the relevance to man of effects produced in animals and in selecting the most suitable species for longer-term animal studies.

Information on the kinetic aspects of absorption, distribution and excretion is also valuable in assessing potential hazards. For example, a substance that is only poorly absorbed from the gut is less likely to reach toxic levels in the tissues than one that is rapidly and extensively absorbed. Similarly, a substance that is rapidly excreted may present less hazard than one only slowly eliminated and which might consequently accumulate in the body during prolonged exposure. In terms of designing animal studies, it is important to recognise that normal metabolic and excretory pathways may become swamped at the high dose levels usually employed in toxicity tests. Effects resulting from this situation need to be distinguished from those reflecting the inherent biological activity of the test substance.

Most of the chemicals to which man is exposed that are foreign to the normal metabolic pathways of the body are metabolised by the so-called drug-metabolising enzyme system located chiefly in the liver. With repeated exposure, many chemicals can markedly increase the activity of these enzymes and in so doing they may affect not only their own metabolism but also the metabolic fate of a wide range of other chemicals to which man may be simultaneously exposed.[7] This can have important toxicological implications and, moreover, since several endogenous substances such as steroid hormones and vitamin D are also substrates of the enzyme system, there may be physiological consequences as well. For these reasons, studies of the effect of chemicals on drug-metabolising enzyme activity are being increasingly advocated.

Short-term Studies
Under this heading come studies of about 21 to 90 days duration in rodent species and up to a year in larger animals. Their aim is to characterise the effects of repeated exposure to the test substance, including the possibility of cumulative action, to determine the approximate dose levels at which any effects occur and to indicate the extent to which further testing is necessary.

Studies of 3 to 4 weeks duration at fairly high dose levels are sometimes used to indicate which organs are the major targets for any toxic effects of the test substance, but the most widely used protocol is the 90-day study in rodents. In this, groups of animals, usually rats, are exposed to the test substance in the diet or drinking water for a period of about 90 days.[3] A range of dose levels is used, ideally extending from one causing no detectable effects to a level producing clearly toxic, but non-lethal, effects.

During the study, animals are observed for their general appearance, behaviour, food intake, growth and mortality. Periodical clinical tests on blood and urine samples are carried out and, sometimes, tests for functional abnormalities in major organs or the immune or endocrine systems are performed. At the end of the exposure period, each animal is subject to a thorough post-mortem examination. The weights of major organs and glands are recorded and tissue samples preserved for microscopic examination.

A properly designed and conducted short-term study should detect most potential toxicities. Using multiple dose levels enables dose–response relationships to be established and, hopefully, will identify a dose level at which no effects occur. This 'no-effect level' will be used in assessing a safe level for human exposure, but usually additional studies will first be required to characterise more fully the biological activity of the test substance.

Reproduction and Teratogenicity Studies

The potential of chemicals to affect reproduction is receiving increasing attention. A wide variety of protocols is available for probing effects at all stages of the reproductive process from gametogenesis to maturation of the offspring.[2] Typically, weanling rats are raised to maturity, mated and the offspring taken through one complete reproductive cycle, with exposure to the test substance being maintained throughout. This design can yield information on mating behaviour, fertility, the progress of pregnancy and the post-partum progress of both mothers and offspring. Studies of this nature are usually conducted as separate investigations where earlier studies (short-term, metabolic, mutagenicity) suggest that reproduction might be affected. However, a short-term test protocol can readily be extended to include a reproduction study with economies of both animals and time.[3]

Chemicals which can cross the placenta may directly affect the development and survival of embryos *in utero*. Such effects may not be fully revealed in the design just mentioned because of the tendency of mothers to cannibalise malformed or dead offspring at birth. By killing the pregnant females just before normal parturition, the contents of the uterus can be examined directly for dead or malformed foetuses. This type of study is the basis of the specialised discipline of teratology.

Long-term Studies

These involve administration of the test substance to animals for the majority of their lifespan with the object of detecting toxic effects that develop only after prolonged exposure. With few exceptions, this means the detection of carcinogenic potential since most other effects will be manifest in earlier studies. Because of their length and complexity, long-term studies are subject to many more potential variables than other forms of toxicity assessment. Many of these variables can affect the outcome of the study and so the experimental design is of crucial importance. Several reviews deal at length with procedures for long-term testing[8,9,10] but there is a lack of general agreement over a number of aspects of design.[3,11]

Among the contentious issues is the choice of animal species and strain. Different species and strains differ considerably in their sensitivity to chemical carcinogens and in their susceptibility to the spontaneous development of tumours unrelated to treatment. Animals therefore need to be selected in the light of as much background information as possible, based on previous experience, and there is no consensus presently as to the most appropriate choice.[2,3] The choice of dose levels, particularly the highest dose, is crucial.[12] A dose that is too high may result in premature death of the animals whereas one too low will reduce the sensitivity of the test. Experience gained in the short-term tests is the guide here. It is sometimes argued that a single dose level (and a control) is adequate to answer the question 'Is the test substance carcinogenic or not?' Although this is the cheapest alternative, such a study may be rendered valueless by inappropriate dose selection. The use of several dose levels reduces this risk and increases the value of the results enormously because it provides the opportunity to establish a dose–response relationship. Mathematical models are now available which can utilise such data to generate a quantitative assessment of human risk.[2,3] A further contentious point concerns the optimum length for a long-term study. Improvements in the quality of laboratory animals mean that many will survive well beyond the 18-month and two-year exposure periods traditionally recommended for rats and mice, respectively. It is argued that the chances of detecting a carcinogenic effect increase with the duration of exposure. However, the spontaneous tumour incidence also increases with time, creating increased background 'noise' against which an induced increase in tumour incidence may be very difficult to establish.

While these and other points continue to be debated, it is currently recommended that long-term studies be carried out in two species of rodent. These are usually rats and mice, a minimum of 50 animals of each sex for each treatment group being exposed for a period of two years. Other species may be used if metabolic studies show major differences between man and the rat or mouse, but practical difficulties place severe restrictions on the use of larger species.

Genetic Toxicology

Growing awareness that one undesirable effect of chemicals may be the production of mutations has established genetic toxicology as the most recent and most rapidly developing aspect of toxicology. 'Spontaneous' mutations have long been known to occur both in man and other organisms. More recently, the demonstration that certain chemicals can cause mutations in bacteria and mammalian cells has raised the possibility that some, at least, of the mutations that occur in man may be due to environmental chemicals. Genetic toxicology is concerned with the detection of mutagenic agents and the possible hazard they pose in terms of inheritable genetic damage.

A wide variety of laboratory techniques for detecting mutational events is now available, largely as a result of advances in our understanding of the molecular basis of mutagenesis, in terms of structural damage to the genetic material, DNA.[13] The multiplicity of tests available reflects not just differing methodologies but the differing molecular mechanisms by which genetic damage can occur such that no single test system could in itself be adequate. The available tests range from *in vitro* systems using bacteria, e.g. the Ames test,[14] fungi and mammalian cells in culture, to *in vivo* tests such as the micronucleus test, the dominant lethal test and tests for chromosome aberrations. The application of these and a wide range of other tests is fully discussed by the Food Safety Council.[3]

Of the chemicals so far examined for mutagenicity, a high proportion of those giving positive results are also known to be carcinogenic from animal studies or human experience. This strong correlation between mutagenicity and carcinogenicity[15] means that mutagenicity tests could be useful also as tests for carcinogenicity. As such, they would have the great advantages of rapidity—a few days to a few weeks—and low cost in comparison with long-term animals studies. The exploitation of mutagenicity tests in this way is therefore being intensively studied. So far, most experience has been accrued with the Ames test,[14] the cell transformation test[16,62]—in which mammalian cells are exposed in culture to the test chemical—and tests based on the measurement of DNA repair as an index of chemically induced damage to DNA.[17]

For these systems, predictive accuracy of 80–90 % has been claimed in validation studies with known carcinogens and non-carcinogens.[17,18] This level of predictability means that false-positive and false-negative results do occur, and at present there is no question of such tests replacing long-term carcinogenicity studies in animals. However, different test systems differ to some extent in the range of carcinogens they can detect, and consequently the use of a battery of several different systems may achieve a better predictive accuracy than a single test.[18] At present, the rapid tests are proving useful as a preliminary screen in the development of new chemicals and in assessing priorities for long-term testing.

Trends in Test Selection

A major influence on toxicological practice over the past decade has been the greatly increased emphasis given to testing for carcinogenicity. At present, long-term tests in animals are still considered the only acceptable means for evaluating carcinogenic potential. But such tests are slow and expensive, and demand already outstrips the facilities available. The decision as to the necessity for long-term testing, or whether a reasonable assurance of safety can be given on the basis of shorter-term tests and considerations of chemical structure and human exposure, is therefore an increasingly critical one.

A significant contribution in this respect has recently come from the Food Safety Council in America.[3] They have organised the available test procedures into a 'decision tree' aimed at providing criteria for decisions on safety to be made at as early a stage as possible. In this sequence, a substance acceptable on the grounds of acute toxicity would next be subject to a battery of mutagenicity tests and to metabolic studies prior to initiating short-term animal studies. The design of the latter would take advantage of the information already generated. The need for long-term testing would be decided primarily on an assessment of genetic toxicity, metabolic fate, the results of the 90-day test and considerations of chemical structure and human exposure levels. Uncertainty in any of these aspects would trigger a decision to proceed. On the other hand, firm evidence, for example, of mutagenicity could lead to rejection of the test substance at that early stage, while favourable results in all of the earlier studies could provide sufficient assurance of safety to obviate the need for long-term testing.

A rather different approach is offered by Cramer et al.[19] A decision tree based on chemical structure and known data on metabolism and toxicity is used to predict the extent of testing required and thereby establish priorities for testing. The use of chemical structure analysis to predict biological activity is becoming an increasingly powerful tool. Recently Jurs et al.[20] described a computer program capable of achieving a predictive accuracy of around 85% in studies with a miscellaneous group of 209 known carcinogens and non-carcinogens. Undoubtedly, developments in both of these areas will have a major impact on toxicity assessment in the future.

INTERPRETATION OF RESULTS

Setting an acceptable level for human exposure to a food additive requires translation of the results of toxicity studies in animals into a quantitative assessment of their implications for human risk. This is one of the most difficult areas of toxicology, for the applicability of such results to man is by no means automatic or direct. The imponderables encountered are numerous and include species differences in sensitivity, the use of experimental dose levels far greater than expected use levels, the possibility of synergistic effects in man resulting from multiple exposure to other chemicals and the fact that laboratory animals are usually a healthy, homogeneous population whereas the human population is highly heterogeneous.

Traditionally, acceptable human intakes[2,21] are based on the highest dose level in animal studies at which no adverse effects due to treatment are evident. A 'safety factor' of 100 is then applied to allow for the uncertainties in extrapolating from animals to man.[2,21] For example, a no-effect level in rats of 100 mg/kg body weight/day would give an acceptable daily intake (ADI) for man of 1 mg/kg, or a total of 60–70 mg for an average adult. This approach to establishing an ADI is still very widely used but it does have limitations.

Safety Factors
A factor of 100 is traditionally used, but this is quite arbitrary and rigid adherence to it is unreasonable. Where an additive is rapidly and completely metabolised to products which are normal body or dietary constituents, a lower factor is justified. A similar case could be made where the results of appropriate studies in man are available, where very low usage levels are envisaged or where the effects seen in animals are considered of marginal significance. Conversely, an increased safety factor may be justified when a compound exerts its toxic effects on an important physiological target or where its use will be in foods whose consumption fluctuates widely, e.g. ice cream, or which are consumed in quantity by sensitive population groups, e.g. children.

The 'No-Effect Level'
Acceptable human intakes are based on no-effect levels established in

animal studies. The predictive value of a no-effect level, however, depends very much on the sensitivity of the procedures used to detect effects. It is also considerably dependent on the number of animals tested, for the statistical implications of 0 out of 10 and 0 out of 100 are very different.

From a practical standpoint, it can often be difficult to decide whether a particular change represents an adverse, i.e. toxic, effect or whether it reflects an adaptive response or some incidental effect of little toxicological significance. A good example of this sort of problem is the interpretation of liver enlargement, an effect frequently encountered in toxicity tests. If the enlarged liver shows histological signs of damage then clearly the effect is a toxic one. More often, the enlarged liver appears histologically normal and the significance of the effect cannot be assessed without further studies. Such studies[22,23] have indicated that an enlarged but histologically normal liver may reflect induction of drug-metabolising enzymes in response to high levels of the test compound. It would be more reasonable to interpret liver enlargement of this nature as an adaptive rather than a toxic response. There are many other examples of changes whose significance may not be immediately clear. Thus, diarrhoea could represent a toxic effect or it could merely be of osmotic origin due to high levels of test compound in the diet. Similarly, animals may lose body weight due to toxic anorexia or simply because incorporation of the test compound makes their diet unpalatable.

The No-Effect Level and Chemical Carcinogens
The concept of a no-effect level implies that there exists a threshold dose for any substance below which no effects will be produced. Whether this concept can be applied to chemical carcinogens, however, is one of the most contentious issues in contemporary toxicology. On the one hand, a relationship between dose and tumour incidence has been established for several carcinogens,[24] raising the possibility of a threshold level below which no tumours will result. On the other, it can be argued that even a single molecule of a carcinogen may be sufficient to initiate the cancer process.[25] Early resolution of this issue is unlikely both because of the experimental difficulties of studying dose–response relationships at very low levels of exposure and because insufficient is known of the mechanisms of chemical carcinogenesis. From a practical standpoint, however, the relationship between dose level and latent period[26] may be relevant in that, even if there is no absolute threshold for carcinogens, there may well be thresholds below which the latent period for the development of tumours exceeds the lifespan of animals or man.[24] These issues are more than dry academic debate. Known carcinogens are present in some unprocessed foods and they may be

formed during processing, one of the best known examples being the formation of nitrosamines in nitrite-preserved fish and meat.

Future Trends

Because of the types of problem mentioned, acceptable daily intakes as traditionally established have little absolute meaning. Moreover, toxicological studies with food additives have overlooked, and still do largely overlook, the problem of interactions between the additive and food components. Such interactions are well known to occur, e.g. with sulphur dioxide and nitrite, and mean that the material ingested by man may bear little resemblance to the pure additive tested in animals. No practical solution to this dilemma is presently in sight.

Despite the difficulties, it may be argued that current procedures work well in that serious adverse effects attributable to food additives are very rare. This situation, however, owes much more to the application of sufficiently large safety factors than to an understanding of the biological activity of food additives. But the use of essentially arbitrary safety factors tends to make the system inflexible so that uncertainties raised by the application of inappropriate or ill-understood experimental methods may result in the unnecessary restriction of potentially valuable substances. A more rational and hence more flexible approach to the interpretation of toxicity data will only be possible with increased understanding of mechanisms of toxicity. This is a long-term goal. In the meantime, increased emphasis on dose–response relationships in toxicity studies offers the brightest hope of a less empirical method of establishing human exposure levels. A key factor here will be the utilisation of mathematical models which allow the extrapolation of experimental dose–response data so that quantitative estimates of response can be made at likely levels of human exposure.[3]

TOXICOLOGY OF PRESERVATIVES

The preservatives currently permitted for food and beverage use in the UK and other countries were documented in Chap. 2. This section briefly reviews the current toxicological status of the major preservatives.

(a) Sulphur Dioxide

Sulphur dioxide and its related anions (sulphite, bisulphite and metabisulphite) have achieved very extensive usage in foods and beverages.

The toxicity of orally administered SO_2 has been widely studied[21] but the present acceptable daily intake (ADI) was established largely on the basis of long-term and multi-generation studies reported by Til *et al.*[27,28] In these, a no-effect level equivalent to 72 mg free SO_2/kg body weight/day was obtained, giving an ADI of 0·7 mg/kg or 45–50 mg/average man/day.[21] It is difficult to estimate an average intake of SO_2 for man since consumption of treated foods varies so widely, but because of its broad utility, SO_2 intake may come close to or even exceed the ADI. For example, the consumption of about three glasses of wine per day could alone lead to an SO_2 intake exceeding the ADI. Other sources of exposure to SO_2, e.g. atmospheric pollution, must also be considered.

The implications of this are difficult to gauge for, as pointed out earlier, an ADI has no absolute meaning and the studies of Til *et al.* did indicate that quite high dietary levels of SO_2 may be innocuous. There are, however, a number of other, more imponderable foci for concern in relation to the biological activity of SO_2. Most of these stem from the unique reactivity of SO_2, both in foods and in living systems.

Thiamine (vitamin B_1) in foodstuffs is destroyed by SO_2, raising the possibility that adverse effects might result from an induced thiamine deficiency. This problem has largely been resolved by studies showing that animals fed nutritionally adequate diets can withstand substantial intakes of SO_2 in terms of thiamine destruction[29] and that humans ingesting up to 200 mg SO_2 per day showed no signs of thiamine deficiency or changes in its urinary excretion.[30] Moreover, SO_2 is not permitted in foods that are major sources of thiamine. However, SO_2 also interacts with folic acid,[31] vitamin K[32] and certain flavines and flavoenzymes.[33] The nutritional and biochemical significance of these interactions has not yet been established.

The possibility that toxic products might result from interactions between SO_2 and dietary components was raised by Bhagat and Lockett.[34] They showed that sulphited animal diets stored for 3–4 months produced stunting of growth and diarrhoea that could not be ameliorated by thiamine supplementation. A wide variety of reaction products have now been identified in SO_2-treated foods. These include carbohydrate sulphonates and hydroxysulphonates, amino acid and peptide S-sulphonates and analogous derivatives of many other food constituents.[35,36] The toxicological significance of these products is essentially unknown. Their presence therefore introduces uncertainties in assessing the safety in use of SO_2, as the nature and amounts of interaction products formed will clearly vary with different food systems and differing conditions of processing and storage.

Sulphur dioxide also reacts with cellular constituents *in vivo* and so could interfere in diverse ways with metabolic processes. The capacity of SO_2 to react with proteins, chiefly via sulphitolysis of disulphide bridges, has potentially important biological implications in view of the role of such linkages in maintaining the structural integrity of proteins. There is no direct evidence bearing on this, although a suggestion has been made that SO_2 may interfere with the immune response by affecting antibody structure.[37]

In addition to proteins, SO_2 also reacts with nucleic acids. A variety of reactions have been demonstrated including conversion of cytosine (which occurs in both DNA and RNA) to uracil (which occurs in RNA only).[38] The importance of such reactions is not clear for, although bisulphite is mutagenic to bacteria,[38] no convincing evidence exists for mutagenicity in mammals. At high concentrations *in vitro*, bisulphite produced growth inhibition and chromosome abnormalities in mammalian cells, but *in vivo* mutagenicity studies have all proved negative.[38] This is probably because of the extensive capacity of mammalian tissues for metabolising SO_2. The possibility that SO_2 poses a genetic hazard for man therefore seems remote.

The capacity of mammalian tissues to metabolise SO_2 is particularly important. The enzyme system responsible, sulphite oxidase (E.C. 1.8.3.1.), oxidises sulphite to sulphate and is present in most mammalian tissues although liver, kidney and heart have the highest levels. Under normal conditions, the capacity of this metabolic system is far in excess of the highest levels of SO_2 likely to be ingested from treated foods. However, there is evidence that congenital deficiency of sulphite oxidase can occur in man,[39] and that in the rat the activity of the enzyme system is very low in the neonate.[40] Certainly it can be shown in animals that the toxicity of sulphite is inversely related to the activity of sulphite oxidase.[41] Further information on the limits of variation of sulphite oxidase activity in man is therefore desirable if SO_2 intake is likely to remain around or in excess of the present ADI.

(b) Sorbic Acid

By comparison with SO_2, the toxicological status of sorbic acid (*trans,-trans*-2,4-hexadienoic acid) is much less contentious. Early work showed sorbic acid to have a very low order of acute toxicity, the oral LD_{50} for rats being about 7–10 g/kg,[21] while in short-term studies dietary levels as high as 10 % were tolerated with only minor effects.[42] Sorbic acid is metabolised via the normal, physiological pathways of fatty acid oxidation, pathways common both to laboratory mammals and man.[43]

Despite this apparently favourable situation, the 1972 review of the UK Preservatives in Foods Regulations recommended against increasing usage of sorbic acid because of the inadequacy of existing long-term toxicity data. This aspect has since been extensively investigated. In a two-year study in rats, Gaunt et al.[44] established a no-effect level of 1·5 % of the diet with only minor effects noted at a dietary level of 10%. There was no evidence of carcinogenicity. However, concern had been expressed at the possibility of sorbic acid being contaminated with para-sorbic acid (5-hydroxy-2-hexenoic acid δ-lactone) or of the latter forming during storage or cooking of sorbic acid-containing foods. Several unsaturated lactones, including para-sorbic acid, have been shown to produce sarcomas at the site of repeated subcutaneous injections.[45] However, the significance of sarcomas produced in this manner, as an index of carcinogenicity for orally ingested substances such as food additives, is very doubtful.[46] In the case of para-sorbic acid, these doubts have been borne out by long-term studies in both rats and mice with sorbic acid deliberately adulterated with 1000 ppm (parts per million) of para-sorbic acid: in neither species was there any evidence of carcinogenicity.[47,48] Furthermore it now appears that commercial sorbic acid contains little, if any, para-sorbic acid.

An ADI of up to 1500 mg for a 60 kg human was established by WHO in 1974[21] and the long-term data published subsequently provide no reason to reduce this. In the USA, the Food and Drug Administration (FDA) have affirmed sorbic acid as GRAS (generally recognised as safe) on the basis of a review of all available toxicological information.[43] A number of new applications for sorbic acid in foods have recently been approved by the UK authorities.[49] Finally, it is worth noting that increased use of sorbic acid, for which the available toxicological data are particularly favourable, could allow some reduction in SO_2 intake in situations where the two preservatives may be technologically interchangeable.

(c) Nitrite and Nitrate

These are usually considered together because of the relative ease with which nitrate may be reduced to nitrite by organisms of the intestinal microflora. Both anions have a wide natural distribution but their use as deliberate food additives, chiefly in cured meats, has become a focal point for toxicological concern.

The acute toxic effects of nitrite, resulting from oxidation of haemoglobin to methaemoglobin, have long been known.[50] Infants of less than six months are particularly susceptible and fatal cases of poisoning in infants and children have been recorded following ingestion of vegetables or water containing high levels (100–500 ppm) of nitrite.[50]

Current concern, however, centres around two different aspects. The first of these relates to the observed interaction of nitrite with various nitrogenous components of foods, particularly amines, to form nitrosamines. Many nitrosamines are known to be potent carcinogens in laboratory animals and so it is clearly important to establish the significance of their presence in foods. The principal nitrosamines found in foods are N-nitrosopyrrolidine, N-nitrosodimethylamine, N-nitrosodiethylamine and N-nitrosopiperidine.[51,61] A recent analysis indicated that the total nitrosamine content of cured meats and fish seldom exceeds 1–2 ppb (parts per billion (10^{12})) although occasional samples of fried bacon contained up to 50–100 ppb of nitrosopyrrolidine.[51] Based on a survey of the normal UK diet, total nitrosamine intake was calculated to be about 4 μg per person per week.[51] Dietary nitrate and nitrite may also act as precursors for nitrosamine formation $in\ vivo$,[52] in the mouth, stomach and urinary bladder. However, compared with the amounts present naturally in food and water and that produced by endogenous synthesis in the body,[53] the amounts of nitrate and nitrite added to food as preservatives are very small.

Assessment of the hazard to human health posed by the presence of low levels of nitrosamines in foods impinges on one of the most difficult questions in toxicology, that of whether no-effect levels exist for carcinogens. Certainly the levels of nitrosamines required to produce tumours in animals, although small, exceed considerably those present in the whole human diet. But it is argued that in the absence of dose–response data at realistically low levels of exposure, no firm conclusions can be drawn. This type of information is currently being sought in animal experiments and a major study with nitrosamines has recently been completed at BIBRA. An alternative approach, involving administration to animals of nitrite-treated diets containing realistic levels of nitrosamines, has so far failed to show any evidence of carcinogenicity under such conditions.[50,54]

The second area of concern relates to the biological activity of nitrite per se. There has been renewed interest in this question as a result of a long-term study conducted for the FDA in which rats were exposed to nitrite in the diet or drinking water at levels from 250 to 2000 ppm. The essential feature of these results was a slightly increased incidence of reticuloendothelial tumors (lymphoma) in the treated animals—12·5% (172/1380 rats) compared with 8·4% (48/573 rats) in the controls. (Unpublished report to FDA. Slightly different figures are given in ref. 55.) If a proliferative lesion in the spleen and lymph nodes was added to the lymphomas, the combined incidence of both lesions was 23·7% compared with 15·3%, a statistically significant increase. The possibility that nitrite may be carcinogenic *per se* has therefore been raised. However, before accepting any such conclusion, several areas of doubt in this study must be considered. These include the unusually high incidence of lymphomas in the control rats, the significance of the relatively small increases in tumour incidence, the lack of any dose–response relationship and the influence of nitrite toxicity at such high dose levels. Thus, the results of this study complicate rather than facilitate the safety evaluation of nitrite.

Current regulatory attitudes to the use of nitrate and nitrite were reviewed in Chap. 2. In the USA, the results of the nitrite study have accelerated moves to reduce and ultimately phase out the use of nitrate/nitrite, moves already in progress both there and in the UK because of continuing uncertainty over the significance of nitrosamine formation. Nevertheless, while minimising added nitrite levels seems prudent, it must be restated that the lethal effects of *Clostridium botulinum* toxin present a certain consumer hazard whereas the hazard posed by the use of nitrate/nitrite as preservatives is, at present, no more than academic. Certainly it is remarkable, if the levels in foods of nitrosamines or nitrite *per se* were toxicologically significant, that no unusual incidence of cancer has been even tentatively linked with high consumption of preserved meats during the 100 years in which nitrites have been added to these foods.[56]

(d) Benzoic Acid and Salts

The use of benzoic acid and its sodium and potassium salts is supported by a large volume of data establishing the low toxicity of these substances in experimental animals and man.[21] In man and in the rat and rabbit, benzoic acid is excreted rapidly almost entirely as its glycine conjugate, hippuric acid.[21] The latter arises normally as a minor product of metabolism of the aromatic amino acids, phenylalanine and tyrosine.

WHO/FAO[21] established an ADI of 350 mg for a 70 kg human on the basis of a no-effect level in rats of 1 % in the diet. Benzoic acid can produce skin sensitisation when applied topically and there have been reports of allergic responses to orally ingested benzoate in sensitive individuals.[57,58] At present, however, these findings provide no grounds for reconsidering the use of benzoates.

(e) Alkyl Esters of *p*-Hydroxybenzoic Acid

As with benzoic acid, ample toxicological data exist to support the use of methyl-, ethyl- and propyl-*p*-hydroxybenzoate[21] although again the possibility of allergic reactions to p-hydroxybenzoic acid—liberated by hydrolysis of the esters in the gut—has been raised.[57]

The heptyl ester exhibits greater anti-microbial activity than the lower members of the series[59] and has been suggested as a possible alternative to SO_2 in beer. It is permitted at up to 20 ppm in soft drinks in the USA but not in the UK or EEC. In the 1972 review of the Preservatives in Foods Regulations, additional toxicological data were called for, particularly in relation to possible tissue storage, if the use of heptyl-p-hydroxybenzoic acid were to be approved in the UK. To date, such data have not been forthcoming.

(f) Propionic Acid and Salts
Propionic acid is a normal, physiological metabolite in man, being a product of the oxidation of certain fatty acids. On this basis, the FAO/WHO[21] did not limit the ADI, although technological limitations restrict the amount and scope of its usage as a food preservative.

(g) Hexamethylenetetramine
Hexamethylenetetramine (hexamine) finds limited application as a preservative in the UK and EEC (see Chap. 2). In the food matrix and *in vivo*, hexamine degrades yielding formaldehyde. The toxicity of hexamine is therefore closely bound up with that of formaldehyde and the *in vivo* oxidation product of the latter, formate. Formate is a normal metabolic intermediate in mammals.

Early studies, in which hexamine produced local sarcomas after repeated subcutaneous injection and formaldehyde produced mutations in insects, raised the possibility that hexamine could be carcinogenic.[21] However, as pointed out earlier, the production of subcutaneous sarcomas by repeated injection is not considered a relevant index of carcinogenicity for food additives.[46] Similarly, the mutations produced in insects probably have little relevance in view of the high levels of formaldehyde used and since virtually no formaldehyde as such can be detected in mammalian body tissues after administration of either formaldehyde or hexamine due to rapid oxidation to formate. These objections have been borne out by subsequent long-term, multi-generation and teratogenicity studies involving dietary administration of hexamine. Reviewing these studies FAO/WHO[21] allocated an ADI of 10 mg for a 70 kg human. An observation that hexamine can interact with nitrite to form a nitrosamine is of little moment in view of the infinitesimal contribution hexamine would make to the total dietary load of amines.

(h) Diethylpyrocarbonate

Diethylpyrocarbonate (DEPC) was developed for use in beverages and appeared in many respects to be an 'ideal' food preservative, having a favourable anti-microbial spectrum, little overt toxicity and being easily degraded to CO_2 and ethyl alcohol. However, studies showing the formation of trace amounts (20–50 ppb) of the carcinogen, urethane (ethylcarbamate), in DEPC-treated wines and soft drinks[60] led to the withdrawal of DEPC.

This case makes an interesting comparison with the controversy over the presence of nitrosamines in nitrite-preserved foods. In both instances, lack of information on the relationship between very-low-level exposure to carcinogens and tumour induction precludes a firmly based risk assessment. Nevertheless, the use of nitrite has continued because of the tangible risk of botulism in its absence, whereas no comparable benefit can be balanced against the possible risks of urethane formation in DEPC-treated beverages.

CONCLUSION

Food preservatives should ideally be capable of inhibiting unwanted biological reactions in foods without interfering with essentially analogous processes in the tissues of those consuming the treated food. This 'selective toxicity' required of preservatives distinguishes them from all other groups of food additives, whose technological function is independent of any biological activity they may possess. The safety of preservatives is frequently questioned on grounds of non-selectivity of their biological activity. However, with two major exceptions, the toxicological status of the currently permitted preservatives is relatively uncontentious. The exceptions are nitrite and SO_2. Nitrite, and particularly the nitrosamines which may be formed from it, are indisputably toxic to mammals but the hazard posed by the amounts of these substances actually present in foods remains to be resolved. In the case of SO_2, its unique reactivity in biological systems has given rise to a wealth of speculation as to its potential toxicity. Further studies are needed to place these doubts on a firmer scientific foundation.

Among the preservatives, nitrite and SO_2 will undoubtedly continue to be the focal point for toxicological scrutiny. It should go without saying, however, that hazards identified must be balanced against the indisputable benefits accruing from the use of chemical preservatives.

ACKNOWLEDGEMENT

I should like to thank Dr Paul Grasso for invaluable discussions during the preparation of this manuscript.

REFERENCES

1. Ministry of Agriculture, Fisheries and Food, (1965) *Memorandum on procedures for submissions on food additives and on methods of toxicity testing*, HMSO, London.
2. WHO (1978) Environmental Health Criteria 6, *Principles and Methods for Evaluating the Toxicity of Chemicals*, Part I, World Health Organisation, Geneva.
3. Food Safety Council (1978) *Fd Cosmet. Toxicol.*, **16**, Suppl. 2.
4. MORRISON, J. K., QUINTON, R. M. and REINERT, H. (1968) In: *Modern Trends in Toxicology*, Vol. 1, E. Boyland and R. Goulding (Eds.), Butterworths, London, pp. 1–17.
5. SPERLING, F. (1976) In: *New Concepts in Safety Evaluation*, M. A. Mehlman, R. E. Shapiro and H. Blumenthal (Eds.), *Advances in Modern Toxicology*, Vol. 1, John Wiley & Sons, New York, pp. 177–91.
6. FRAZER, A. (1970) In: *Metabolic Aspects of Food Safety*, F. J. C. Roe (Ed.), Blackwell Scientific Publications, Oxford, pp. 1–15.
7. PARKE, D. V. (1975) In: *Enzyme Induction*, D. V. Parke (Ed.), Plenum Press, London, pp. 207–72.
8. GOLDBERG, L. (Ed.) (1973) *Carcinogenesis testing of chemicals*, CRC Press, Cleveland, Ohio.
9. SONTAG, J. M., PAGE, N. P. and SAFIOTI, V. (1976) *Guidelines for carcinogen bioassay in small rodents*, National Cancer Institute Carcinogenesis Technical Report Series, No. 1, DHEW Publication No. NIH 76 801, Washington, DC.
10. PAGE, N. P. (1977) *J. Environ. Path. Toxicol.*, **1**, 161.
11. GRASSO, P. (1979) *Chem. Ind.* (London), 73.
12. MUNRO, I. C. (1977) *J. Environ. Path. Toxicol.*, **1**, 183.
13. BRIDGES, B. A. (1979) In: *Mutagenesis in Sub-Mammalian Systems*, G. E. Paget, (Ed.), *MTP Press*, Lancaster, pp. 3–12.
14. AMES, B. N., McCANN, J. and YAMASAKI, E. (1975) *Mutat. Res.*, **31**, 347.
15. McCANN, J. and AMES, B. N. (1976) *Proc. Natl Acad. Sci. USA*, **73**, 950.
16. STYLES, J. A. (1977) *Br. J. Cancer*, **36**, 558.
17. MARTIN, C. N., McDERMID, A. C. and GARNER, R. C. (1978) *Cancer Res.*, **38**, 2621.
18. PURCHASE, I. F. H., LONGSTAFF, E., ASHBY, J., STYLES, J. A., ANDERSON, D., LEFEVRE, P. A. and WESTWOOD, F. R. (1978) *Br. J. Cancer*, **37**, 873.
19. CRAMER, G. M., FORD, R. A. and HALL, R. L. (1978) *Fd Cosmet. Toxicol.*, **16**, 255.
20. JURS, P. C., CHOU, J. T. and YUAN, M. (1979) *J. Med. Chem.*, **22**, 476.
21. FAO/WHO Joint Expert Committee on Food Additives—17th Report (1974) WHO Techn. Rep. Ser., No. 539, World Health Organisation, Geneva.

22. GRASSO, P., WRIGHT, M. G., GANGOLLI, S. D. and HENDY, R. J. (1974) *Fd Cosmet. Toxicol.*, **12**, 341.
23. CRAMPTON, R. F., GRAY, T. J. B., GRASSO, P. and PARKE, D. V. (1977) *Toxicology*, **7**, 289.
24. JONES, H. B. AND GRENDON, A. (1975) *Fd Cosmet. Toxicol.*, **13**, 251.
25. CRUMP, K. S., HOEL, D. G., LANGLEY, C. H. and PETO, R. (1976) *Cancer Res.*, **36**, 2973.
26. DRUCKREY, H. (1967) In: *Potential Carcinogenic Hazards from Drugs. Evaluation of Risks*, R. Truhaut (Ed.), UICC Monograph Series, Vol. 7, Springer Verlag, Berlin, pp. 60–78.
27. TIL, H. P., FERON, V. J. and DE GROOT, A. P. (1972) *Fd Cosmet. Toxicol.*, **10**, 291.
28. TIL, H. P., FERON, V. J., DE GROOT, A. P. and VAN DER WAL, P. (1972) *Fd Cosmet. Toxicol.*, **10**, 463.
29. HOTZEL, D., MUSKAT, E., BITSCH, I., AIGN, W., ALTHOFF, J. D. and CREMER, H. D. (1969) *Int. Z. Vitaminforsch.*, **39**, 372.
30. CREMER, H. D. and HOTZEL, D. (1970) *Int. Z. Vitaminforsch.*, **40**, 52.
31. VONDERSCHMITT, D. J., VITOLS, K. S., HUENNEKENS, F. M. and SCRIMGEOUR, K. G. (1967) *Arch. Biochem. Biophys.*, **122**, 488.
32. SHIH, N. T. and PETERING, D. H. (1973) *Biochem. Biophys. Res. Commun.*, **55**, 1319.
33. HEVESI, L. and BRUICE, T. C., (1973) *Biochemistry*, **12**, 290.
34. BHAGAT, B. and LOCKET, M. F. (1964) *Fd Cosmet. Toxicol.*, **2**, 1.
35. SCHROETER, L. C. (1966) *Sulphur Dioxide: Applications in Foods, Beverages and Pharmaceuticals*, Pergamon Press, Oxford.
36. GILBERT, J. and MCWEENY, D. J. (1976) *J. Sci. Fd Agric.*, **27**, 1146.
37. ZARKOWER, A. (1972) *Arch. Environ. Health*, **25**, 45.
38. SHAPIRO, R. (1977) *Mutat. Res.*, **39**, 149.
39. SHIH, V. E. *et al.* (1977) *New Engl. J. Med.*, **297**, 1022.
40. JOSHI, V. C., KURUP, C. K. R. and RAMSARMA, T. (1969) *J. Biochem.*, **111**, 297.
41. TEJNOROVA, I. (1978). *Toxicol. Appl. Pharmacol.*, **44**, 251.
42. DEMAREE, G. E., SJOGREN, D. W., MCCASHLAND, B. W. and COSGROVE, F. P. (1955) *J. Am. Pharm. Ass.*, **44**, 619.
43. Food and Drug Administration, (1973) *Scientific literature reviews on generally recognised as safe food ingredients—sorbic acid and derivatives*, National Technical Information Service, US Dept of Commerce, Springfield, Va.
44. GAUNT, I. F., BUTTERWORTH, K. R., HARDY, J. and GANGOLLI, S. D. (1975) *Fd Cosmet. Toxicol.*, **13**, 31.
45. DICKENS, F. and JONES, H. E. H. (1963) *Br. J. Cancer*, **17**, 100.
46. GANGOLLI, S. D., GRASSO, P., GOLDBERG, L. and HOOSON, J. (1972) *Fd Cosmet. Toxicol.*, **10**, 449.
47. MASON, P. L., GAUNT, I. F., HARDY, J., KISS, I. S., BUTTERWORTH, K. R. and GANGOLLI, S. D. (1976) *Fd Cosmet. Toxicol.*, **14**, 387.
48. MASON, P. L., GAUNT, I. F., HARDY, J., KISS, I. S. and BUTTERWORTH, K. R. (1976) *Fd Cosmet. Toxicol.*, **14**, 395.
49. Food Additives and Contaminants Committee Report on the Use of Sorbic Acid in Food, (1977) FAC/REP/24, HMSO, London.
50. FAO/WHO Joint Expert Committee on Food Additives—20th Report (1976) WHO Techn. Rep. Ser., No. 599, World Health Organisation, Geneva.

51. GOUGH, T. A., WEBB, K. S. and COLEMAN, R. F. (1978) *Nature*, **272**, 161.
52. RUDDELL, W. S. J., BLENDIS, L. M. and WALTERS, C. L. (1977) *Gut*, **18**, 73.
53. TANNENBAUM, S. R., FETT, D., YOUNG, V. R., LAND, P. D. and BRUCE, W. R. (1978) *Science*, **200**, 1487.
54. KNUDSON, I. and MEYER, O. A. (1977) *Mutat. Res.*, **56**, 177.
55. NEWBERNE, P. M. (1979) *Science*, **204**, 1079.
56. BARNES, J. M. (1974) *Essays Toxicol.*, **5**, 1.
57. MICHAELSSON, G. and JUHLIN, L. (1973) *Br. J. Dermatol.*, **88**, 525.
58. FREEDMAN, B. J. (1977) *Clin. Allergy*, **7**, 405.
59. STRANDSKOV, F. B., ZILIOTTO, H. L., BRESCIA, J. A. and BOCKELMANN, J. B. (1965) *Am. Soc. Brew. Chem.*, Proc. Annu. Meet., 1965, p. 129.
60. SOLYMOSY, F., ANTONI, F. and FEDORCSAK, I. (1978) *J. Agric. Fd Chem.*, **26**, 500.
61. WHO International Agency for Research on Cancer (1978) IARC Monographs on the Evaluation of Carcinogenic Risk of Chemicals to Humans, Vol. 17, *Some N-nitroso compounds*, IARC, Lyon.
62. PIENTA, R. J., POILEY, J. A. and LEBHERZ, W. B. (1977) *Int. J. Cancer*, **19**, 642.

Chapter 4

DETECTION AND ANALYSIS

R. Sawyer and N. T. Crosby

Laboratory of the Government Chemist, London, UK

INTRODUCTION

This review of techniques for the detection and estimation of food preservatives is divided into two parts; inorganic preservatives, and organic preservatives. A number of traditional methods are included since they form the nucleus of accepted official methods. However, the review has concentrated on the main developments in the last decade. In the latter sense, it is a supplement to the review by Schuller and Veen which concentrated on method developments in the 1960s.

PART 1
INORGANIC PRESERVATIVES

SULPHUR DIOXIDE

Sulphur dioxide may be used as a preservative in the form of a gas, or in solution as sulphurous acid, or as the sodium, potassium or calcium salts of the acid. Part of the added sulphite may be fixed subsequently by naturally occurring constituents of food such as aldehydes, ketones, simple sugars, dehydroascorbic acid, thiamine or polyphenols. There may, therefore, be a need to determine both 'bound' and 'free' sulphite in food, although the regulations state a maximum allowable figure which relates to the *total* amount present (bound and free) calculated as SO_2 (w/w). The bound form

75

of sulphite in foods can be released by treatment with alkali, followed by acidification. Alternatively, methods involving distillation also determine total sulphite, although with some foods SO_2 is liberated only with difficulty. Whilst SO_2 is the most commonly used preservative, it is only the undissociated fraction of the unbound sulphurous acid that is effective against micro-organisms.[1] The undissociated fraction is fixed, however, by the natural pH value of the substrate. At pH values in the physiological range, a mixture of sulphite and bisulphite ions is present. Lower pH values increase the proportion of bisulphite ions.

Qualitative Tests

Simple tests for the presence of sulphites are based on the ability of SO_2 to decolorise either malachite green,[2] or para-rosaniline.[3] Alternatively, SO_2 may be reduced by zinc and acid. The hydrogen sulphide liberated is then detected by lead acetate paper, or by potassium iodide and starch. Sulphides, present in some vegetables, will interfere. In the method recommended by Parkes,[4] the sample with acid and marble chips is placed in a conical flask fitted with a bent thistle funnel containing a few drops of iodine solution and barium chloride as a seal. On warming, if sulphites are present, the colour of the iodine is discharged and a white opalescence ($BaSO_4$) is observed immediately the first drop of condensate passes into the thistle funnel.

In all cases, confirmation using a quantitative method is desirable.

Quantitative Methods

Since SO_2 is readily lost from samples on storage, it is essential that the samples are kept in airtight containers and examined as soon as possible after sampling. Jennings et al.[5] showed that for samples of dried peas and lemon juice, the SO_2 content fell to 60 % of its original value in about two weeks at ambient temperature. With dehydrated potato powder, Hearne and Tapsfield[6] found that the effect varied with the moisture content of the product and the atmosphere. Stadtman et al.[7] noticed a similar effect with dried apricots and they showed that SO_2 disappearance was a function of storage temperature. Standard solutions of sulphite also deteriorate on storage at room temperature. Phillips[8] investigated the stability of a 0·1 % solution of potassium metabisulphite at 60 °F and found that it decreased as shown.

Time (h)	0	24	72	96	120	144
SO_2 content (%)	100	98·2	94·6	93·5	92·2	91·3

More dilute solutions deteriorate even more rapidly. Buckee and Hargitt[9] found that an aqueous solution containing initially 20 mg/l of SO_2 had been reduced to 15 mg/l over a period of only 3 h.

Quantitative methods for SO_2 can be divided into two groups: (a) direct determination without distillation, and (b) distillation followed by determination of SO_2 in the distillate. Methods have been reviewed by Schuller and Veen.[10]

(a) Direct Methods

These methods are widely used for quality-control purposes and generally involve titration with iodine. Although the method is simple and fairly rapid, it is unsuitable for coloured or turbid samples. In addition, other substances, e.g. starch, may be present which also react with iodine. This leads to drifting end-points and high blank titres. Combined sulphite is first released by treatment with alkali. The solution is then acidified and titrated with standard iodine solution.[11,12]

An alternative technique involves titration with iodine before, and after, addition of a carbonyl compound. The difference between the two titres represents the SO_2 content.[13] The same authors also report a desorption technique using oxygen-free nitrogen. The SO_2 released is trapped in alkaline glycerol and again estimated using iodine solution. Lower levels (~ 50 mg/kg) are trapped in sodium tetrachloromercurate reagent and then estimated colorimetrically. A microdiffusion technique has been recommended for alcoholic beverages by Owades and Dono[14] but, inevitably, long analysis times are required.

Electrochemical techniques for the determination of SO_2 have also been described.[15] Differential electrolytic potentiometry was used for end-point location down to 10^{-6} molar. However, the technique was not applied to the analysis of foods. Recent interest in electrochemical methods has centred around the gas-sensing electrode probe. The Electronic Instruments Ltd Model 8010-200 probe has been described by Bailey and Riley.[16] This technique is ideally suited to quality-control measurements on a wide range of foods as the response is rapid except at low concentrations. Acetic acid is the most serious interfering species likely to be encountered in foods, e.g. pickles. A comparison of the use of the electrode with two other techniques has been published by Jennings et al.[5]

Molecular emission cavity analysis (MECA) is particularly suitable for the analysis of compounds containing sulphur and the use of the technique for solutions of sulphur dioxide has been described by Belcher et al.[17] In future years this may well prove to be useful as an alternative rapid method for quality-control work.

(b) *Distillation Methods*

Distillation is essential for accurate results with most foods but the method is tedious when large numbers of samples have to be examined. The principle of the method is that SO_2 is released by heating in acidic solution and is then determined in the distillate after oxidation. Errors from sulphides and volatile acids may still occur unless suitable precautions are taken. The basic method was described by Monier-Williams[18] in 1927 but a number of modifications to the apparatus and reagents have been suggested since that time. Both hydrochloric acid and phosphoric acid have been recommended for acidification. The former gives higher values and releases SO_2 from the food more rapidly, but it is less reproducible and is thought to liberate volatile sulphur compounds from foods containing proteins. Most workers prefer to carry out the distillation in an inert atmosphere to reduce the risk of oxidation prior to distillation. Carbon dioxide and nitrogen have been used for this purpose and to remove dissolved oxygen from the solution before heating. Nitrogen is preferable to carbon dioxide as it does not interfere in the subsequent alkalimetric end-point determination. In the Tanner modification[19] methanol is added to the distillation flask to lower the boiling point of the mixture and so reduce the quantity of other volatiles in the distillate. The reflux condenser also helps in this respect and a recent study[9] suggests that the design of the apparatus is critical. The distillation should be completed as rapidly as possible, 10 min being sufficient for many foods and improved recoveries of SO_2 are obtained by turning off the cooling water at the end to allow the hot vapour to flush out the apparatus.[9]

Sulphur dioxide present in the distillate can then be estimated iodometrically, alkalimetrically, gravimetrically or colorimetrically following oxidation to sulphate. Both iodine and hydrogen peroxide have been used for the oxidation stage, the latter being preferred except where iodine is used as combined titrant and oxidising agent. Titration with standard alkali solution will measure all acidic substances present; where interference from other volatile acids is suspected, the result should be confirmed by addition of excess of barium chloride solution. After standing overnight, the precipitate is filtered, ignited and weighed. Alternatively, the excess barium in solution can be estimated by titration using EDTA solution.

Lower levels of sulphur dioxide are usually estimated colorimetrically, the most favoured reaction being the decolorisation of *para*-rosaniline.[20] This method gives a linear calibration curve up to 80 μg of sulphur dioxide.

Collaborative Studies

Methods of analysis for sulphur dioxide have been reviewed by Schuller and Veen[10] but few comparison studies have been published. Jennings et al.[5] examined a wide range of foods and compared the Tanner method with direct determination by electrode and an automatic colorimetric procedure based on the rosaniline reagent. They found that the Tanner method was applicable to most types of sample whereas the electrode was not applicable to certain samples and gave more variable results. However, its use was recommended for control purposes. The automatic method was preferable for the rapid analysis of large numbers of samples.

Buckee and Hargitt[9] compared a modified Monier-Williams procedure recommended by the Institute of Brewing with the Tanner method. The latter method gave slightly lower results and this was ascribed to the design and use of the apparatus employed in the method.

Nury and Bolin[21] compared the official AOAC method with the rosaniline colorimetric method using samples of dried fruits. Similar reproducibility was achieved but the colorimetric method was found to be simpler and less time-consuming. The rosaniline method has also been adopted by the International Committee for Uniform Methods for Sugar Analysis for the determination of SO_2 in white sugar.[22] Following investigations by a UK study group in which the method was compared with the DTNB reagent [5,5-dithio-bis-(2-nitrobenzoic acid)], the rosaniline method was found to be more precise, although marginally less rapid. For molasses, no colorimetric methods were suitable and distillation using the Monier-Williams or Tanner methods was recommended.[22]

NITRATE AND NITRITE

Nitrates and nitrites have been used as preservatives for many years. In addition to their role as curing agents, they are also thought to influence the colour and flavour of the final product. The development of the nitrosamine problem has led to increased interest in the nitrate and nitrite levels occurring in a wide range of foods. Both compounds are widely distributed throughout the environment and, hence, there is little use for sensitive qualitative tests in a modern food laboratory.

Quantitative methods have been developed and investigated extensively for both nitrates and nitrites, in meat products in particular. Methods have been reviewed previously by Schuller and Veen,[10] and by Usher and Telling.[23] The latter review compares critically the published techniques of extraction from meat products, the performance of alternative clearing agents and errors arising from the presence of ascorbate, phosphate, sulphite and the 'meat blank'. Whilst nitrite can be readily and accurately determined in most foods, nitrate methods are subject to a much greater degree of variation and error.

Extraction and Clearing

Nitrates and nitrites are readily soluble in water. Nevertheless, a variety of extraction conditions have been recommended for the extraction stage, including the use of both hot and cold water, as well as neutral or alkaline buffers. Homogenisation increases the efficiency of extraction, otherwise more than one extraction may be necessary. The pH of the extractant should be greater than 5 to avoid loss of nitrite and, in subsequent deproteinisation stages, the pH of the solution should be at a value close to both the isoelectric point of the soluble protein (for maximum precipitation) and as near as possible to the value which leaves the minimum of metal ions in solution, so avoiding hazes at later stages of the determination. The optimum pH range for the precipitation of soluble proteins lies between 5·5 and 6·5.[24] Of the many clearing reagents available, Carrez solutions (zinc acetate and potassium ferrocyanide) are preferred by many workers. Adriaanse and Robbers[25] found that more satisfactory results were obtained when zinc acetate was added first, to prevent any reaction between ferrocyanide and nitrite. Usher and Telling[23] obtained slightly higher recoveries of nitrite on addition of caustic soda to Carrez solutions. They also found that Carrez solutions were more effective protein precipitants than alumina cream, potash alum, or zinc sulphate with borax. Sample preparation procedures for the determination of nitrite in frankfurters have been compared by Fiddler and Fox.[26] The AOAC recommended method involving heat digestion was found to be superior to 11 other treatments examined in which a wide variety of clearing agents was used.

Determination of Nitrite

The diazotisation reaction of nitrous acid forms the basis of most quantitative methods for the determination of nitrite. Whilst only a few different amines have been used to form the diazonium compound, there are a number of alternative amino or hydroxyl compounds that can be coupled to form the azo colour.[27] Despite the highly specific nature of the reaction, conditions such as concentration and nature of reagents, pH value of the solution and temperature, all may affect the rate of the reaction, producing different colour intensities in the final solution. Close attention to the details of the experimental procedure is, therefore, essential to obtain reproducible and accurate results.

Most methods employ sulphanilic acid, or sulphanilamide, for the production of the diazo compound. 1-Naphthylamine, originally used as coupling reagent, has recently been discarded on grounds of carcinogenicity to be replaced either by Cleve's acid (1-naphthylamine-7-sulphonic acid) or by N-1-naphthylethylenediamine (NED). The former reagent[25,28] produces an identical colour and, where nessleriser discs are used to estimate the colour formed, the original discs can still be used with the new reagent. A study of some experimental parameters influencing the rate of colour development in this reaction has been reported by Bunton et al.[29] NED reagent[30] is preferred by some workers as it reacts more quickly and is less variable in composition. Some newer spectrophotometric reagents for the determination of low levels of nitrite in water have been reported by Celardin et al.,[31] whilst fluorometric methods have been tested by Coppola et al.,[32] Axelrod and Engel,[33] Wiersma[34] and by Rammell and Joerin.[35] Fluorometric methods are often more time-consuming but they are particularly suitable for the determination of very low levels of nitrite.

Determination of Nitrate
Spectrophotometric methods for nitrate generally depend on three different reactions:

(a) reduction of nitrate to nitrite or to ammonia;
(b) use of nitrate for the nitration of aromatic compounds in the presence of concentrated sulphuric acid; and
(c) use of nitrate for the oxidation of an organic compound.

In contrast to the determination of nitrite, none of these reactions is specific for nitrate and many interferences have been encountered, particularly at the relatively low levels present in foodstuffs. For example, dissolved organic matter in the presence of strong acids will interfere by competition for nitrate and may also produce charring. Nitrite and chloride are also known to interfere in many of these methods. The nitration of compounds such as 2,4-, 2,6-, or 3,4-xylenol,phenoldisulphonic acid or chromotropic acid is subject to large systematic and random errors. Attempts have been made to improve the methods by distillation or solvent extraction and by the use of masking agents.[10] Oxidation of compounds such as brucine, diphenylamine, strychnine, 1-aminopyrene and salicylic acid generally proceeds erratically. A new calibration curve must be produced for each batch of samples tested. Such methods are generally tedious, particularly when large numbers of samples have to be analysed.

Methods based on the reduction of nitrate to either nitrite or to ammonia have been studied by many workers. Following reduction to nitrite, any of the colorimetric procedures discussed in the previous section can then be used to complete the analysis. Complete reduction of nitrate to ammonia by ferrous sulphate,[36] titanous sulphate,[37] chromous sulphate,[38] or using Devarda's alloy in alkaline solution has not been extensively used in food analysis but has been successfully employed in the water industry.[39] Ammonia produced can then be estimated by nesslerisation, or colorimetrically using the indophenol blue reaction.[40] The difficulty in the use of methods based on the reduction of nitrate is to ensure that reduction to nitrite is complete without any further reduction to ammonia taking place. Various reductants[10,23] such as zinc, copper, cadmium, hydrazine sulphate and micro-organisms have been suggested, but methods based on the use of spongy cadmium are now most widely recommended.[24]

The factors governing the reduction process have been systematically studied by Nydahl.[41] He observed the effects of flow rate, pH, temperature, type of cadmium and buffer used on yield of nitrite. Maximum recoveries were obtained at pH 9·5 where lower flow rates (though less critical) had to be used. There is a need for adequate buffering capacity which also reduces the effect of dissolved oxygen in the solutions. Spongy cadmium, amalgamated cadmium, or cadmium filings, were preferred to other forms of mixed metal reductant in which galvanic cells with higher reducing powers are produced. This conclusion was also supported by Davison and Woof[42] in their study of the batchwise treatment of samples for the reduction of nitrate to nitrite. Follett and Ratcliff[24] also concluded that the efficiency of the cadmium column reaches a maximum in the range pH 9·5–9·7 and they recommended an ammonia buffer solution. However, during the subsequent analysis, the column is eluted with distilled water and not buffer solution so that deviations from the optimum pH value may well occur.

The chief advantage of reduction methods for nitrate is that both nitrate and nitrite can be determined simultaneously in the same extract using the same reagent. Whilst phosphate, chloride and sugar do not interfere, even when present at relatively high concentrations,[24] errors can arise from other sources. Although the reagents produce a small 'blank' value, this is constant and readily corrected for. However, Grau and Mirna[43] showed that some meat extracts contain substances that, after passage through the cadmium column, produce a colour reaction with the nitrite reagents.

Follet and Ratcliff[24] found that the value of the meat blank varied between 13 and 30 mg/kg expressed as sodium nitrate. They suggested that the problem could be overcome by passing a portion of the deproteinised extract, acidified to pH 2·0, through a column of an anion-exchange resin (Deacidite FF). A direct determination of the meat blank can then be made. Ascorbates, when present, yield lower recoveries of added nitrite,[23] since they will react with free nitrite. Ascorbates can be removed by using activated charcoal,[44] or by oxidation with potassium permanganate.[23] The dilution effect in which different ratios of sample to extract are taken has been investigated by Kamm et al.[45] and by Nicholas and Fox.[46] The nitrite response varies non-linearly with the dilution. This effect is thought to be caused by increasing amounts of dissolved oxygen present in the water reacting with substances present in the meat that interfere in the subsequent colorimetric determination. Fiddler and Gentilcore[47] have also shown that some brands of filter paper can contribute significant quantities of nitrite and that the nitrite content can vary from sheet to sheet in the same box.

Nicholas and Fox[46] have reported a critical evaluation of the AOAC method for nitrite including a comparison of four alternative methods. As a result, they made a number of suggestions for the modification of the official method. A recent collaborative study of the colorimetric determination of nitrate and nitrite in cheese has been reported by Hamilton.[48] Clarification of cheese extracts was effected using zinc hydroxide and nitrate was reduced to nitrite with spongy cadmium. The samples contained sodium nitrate at 0, 56, 160 and 280 mg/kg levels. Reasonable agreement amongst the six collaborators was achieved.

Electrochemical Methods
Whilst polarographic methods for the determination of nitrite and nitrate have been known for many years,[49] it is only since the development of ion-selective electrodes that they have been used to any great extent in the food industry.[50] The attraction of this technique is its speed and simplicity especially for routine monitoring or rapid screening applications. Sample preparation prior to the determination of nitrate is restricted to water extraction followed by removal of chloride with silver sulphate and addition of sulphamic acid to destroy nitrite. Determination of nitrate is then achieved by direct immersion of the ion probe. A method of standard additions can be employed for greater accuracy where variations in ionic activities are encountered. The use of the electrode for the rapid screening of baby foods has been described by Pfeiffer and Smith,[51] and by Liedtke and Meloan.[52] In the former study, the electrode was evaluated against the

standard AOAC xylenol method. Forty-nine samples containing nitrate in the range 30–350 mg/kg were examined. A reasonable correlation ($r = 0.91$) was achieved with a standard error of 4·3 mg/kg at a level of 100 mg/kg. At very low levels of nitrate (< 2 mg/kg) the response of the electrode was very slow and unsteady. Cationic resins (Al and Ag) were used to eliminate certain interfering substances. Liedtke and Meloan[52] used a similar approach on a wide range of samples of baby foods. They found that the electrode method gave results for nitrate that were slightly higher than determinations made by the method of Kamm et al.[45]

Other Methods

Automated sampling and analysis methods for nitrite are based on some variation of the diazo colour reaction. Other systems have been proposed for the nitrate determination including enzymatic methods using nitrate reductase.[53–57] More recently, an electrochemical detection method with a dual enzyme system has also been suggested.[58] Unfortunately, this system is not, at present, applicable to very low levels of nitrite or nitrate.

High-performance liquid chromatography has been used for the determination of nitrate by Davenport and Johnson.[59] A strongly basic anion-exchange resin and a cadmium coulometric detector were used. Detection limits as low as 0·1 mg/kg (as N) were achieved and the analysis was completed in less than 10 min.

BORATES

Although borates are no longer permitted preservatives in the UK, their use is allowed in other countries and some textbooks devote space to analytical methods for their detection and analysis. Traces of boron occur naturally in many plant foods and in water.[60]

Qualitative Tests

Most tests for boron depend on the reaction with curcumin. In the AOAC Official Methods book,[61] a preliminary test, in which the food (in fluid form) is acidified and then tested by immersion of turmeric paper (containing curcumin), is described. On drying the paper, a characteristic red colour is obtained if borates are present in the food. Addition of ammonia produces a dark blue/green colour which can be restored with acid. Further confirmation can be achieved by taking another portion of the food, addition of lime-water and evaporation to dryness on a steam bath.

The residue is then ignited at a low red heat until the organic matter chars. After digestion with water and acidification, the test with turmeric paper is repeated. The same book describes a semi-quantitative form of the test for application to meat extracts[62] and an alternative quantitative test for the detection of boron in caviar.[63] Similar tests have been described by Pearson.[64]

Quantitative Methods

The determination of boron can be made titrimetrically, spectrophotometrically, or by using atomic absorption spectrophotometry. The titrimetric procedure involves titration to methyl red or methyl orange followed by the addition of mannitol or glycerol which converts the boric acid to a relatively strong monobasic acid. This is estimated by a further titration to a permanent pink colour using phenolphthalein as indicator. Preliminary treatment of the foodstuff involves ashing in the presence of an alkali or, with some dairy products, more rapid methods using an extracted aqueous phase have been used.[64] Colorimetric procedures are based on the reaction with curcumin[65] in which the absorbance is recorded at both 555 and 700 nm. The boric acid content is then calculated by reference to ΔA (A_{555} − A_{700}) values obtained using standard solutions. Determination by atomic absorption spectrophotometry involves a preliminary wet ashing treatment with nitric and sulphuric acids, followed by extraction using 2-ethyl-1,3-hexanediol in methylisobutyl ketone.

REFERENCES

1. INGRAM, M. (1945) *JSCI*, **67**, 18.
2. HORWITZ, W. (Ed.) (1975) *Official Methods of Analysis*, 12th edn, Association of Official Analytical Chemists, Washington, DC.
3. SCHMIDT, H. (1960) *Z. Anal. Chem.*, **178**, 173.
4. PARKES, A. E. (1926) *Analyst*, **51**, 620.
5. JENNINGS, N., BUNTON, N. G., CROSBY, N. T. and ALLISTON, T. G. (1978) *J. Assoc. Publ. Anal.*, **16**, 59.
6. HEARNE, J. F. and TAPSFIELD, D. (1956) *J. Sci. Fd Agric.*, **7**, 210.
7. STADTMAN, E. R., BARKER, H. A., HAAS, V. and MRAK, E. M. (1946) *Ind. Eng. Chem. (Indust.)* **38**, 541.
8. PHILLIPS, R. J. (1928) *Analyst*, **53**, 150.
9. BUCKEE, G. K. and HARGITT, R. (1976) *J. Inst. Brew.*, **82**, 290.
10. SCHULLER, P. L. and VEEN, E. (1967) *JAOAC*, **50**, 1127.
11. POTTER, E. F. (1954) *Food Technol., Champaign*, **8**, 269.
12. JENSEN, H. R. (1928) *Analyst*, **53**, 133.

13. LLOYD, W. J. W. and COWLE, B. C. (1963) *Analyst*, **88**, 394.
14. OWADES, J. L. and DONO, J. M. (1967) *JAOAC*, **50**, 307.
15. BAILEY, P. L. and BISHOP, E. (1972) *Analyst*, **97**, 311.
16. BAILEY, P. L. and RILEY, M. (1975) *Analyst*, **100**, 145.
17. BELCHER, R., BOGDANSKI, S. L. and TOWNSHEND, A. (1973) *Anal. Chim. Acta*, **67**, 1.
18. MONIER-WILLIAMS, G. W. (1927) *Analyst*, **52**, 343 and 415.
19. TANNER, H. (1963) *Mitt. Geb. Lebensmitt. u. Hyg.*, **54**, 158.
20. KING, H. G. C. and PRUDEN, G. (1969) *Analyst*, **94**, 43.
21. NURY, F. S. and BOLIN, H. R. (1965) *JAOAC*, **48**, 796.
22. Rep. Proc. 16th Session held in Ankara, 2–7 June 1974, I.C.U.M.S.A., p. 226.
23. USHER, C. D. and TELLING, G. M. (1975) *J. Sci. Fd Agric.*, **26**, 1793.
24. FOLLETT, M. J. AND RATCLIFF, P. W. (1963) *J. Sci. Fd Agric.*, **14**, 138.
25. ADRIAANSE, A. and ROBBERS, J. E. (1969) *J. Sci. Fd Agric.*, **20**, 321.
26. FIDDLER, R. N. and FOX, J. B. (1978) *J. Assoc. Offic. Anal. Chem.*, **61**, 1063.
27. SAWICKI, E., STANLEY, T. W., PFAFF, J and D'AMICO, A. (1963) *Talanta*, **10**, 641.
28. CROSBY, N. T. (1967) *Proc. Soc. Wat. Treat. Exam.*, **16**, 51.
29. BUNTON, N. G., CROSBY, N. T. and PATTERSON, S. J. (1969) *Analyst*, **94**, 585.
30. SHINN, M. (1941) *Ind. Eng. Chem. Anal. Ed.*, **13**, 33.
31. CELARDIN, F., MARCANTONATOS, M. and MONNIER, D. (1974) *Anal. Chim. Acta*, **68**, 61.
32. COPPOLA, E. D., WICKROSKI, A. F. and HANNA, J. G. (1975) *JAOAC*, **58**, 469.
33. AXELROD, H. D. and ENGEL, N. A. (1975) *Anal. Chem.*, **47**, 922.
34. WIERSMA, J. H. (1970) *Anal. Lett.*, **3**, 123.
35. RAMMELL, C. G. and JOERIN, M. M. (1972) *J. Dairy Res.*, **39**, 89.
36. PAPPENHAGEN, J. M. (1958) *Anal. Chem.*, **30**, 282.
37. CRESSER, M. S. (1977) *Analyst*, **102**, 99.
38. LINGANE, J. J. and PECSOK, B. L. (1949) *Anal. Chem.*, **21**, 622.
39. Department of the Environment, (1972) *Analysis of Raw, Potable and Waste Waters*, HMSO, London, p. 150.
40. TETLOW, J. A. and WILSON, A. L. (1964) *Analyst*, **89**, 453.
41. NYDAHL, F. (1976) *Talanta*, **23**, 349.
42. DAVISON, W. and WOOF, C. (1978) *Analyst*, **103**, 403.
43. GRAU, R. and MIRNA, A. (1957) *Z. Anal. Chem.*, **158**, 182.
44. FUDGE, R. and TRUMAN, R. W. (1973) *J. Assoc. Publ. Anal.*, **11**, 19.
45. KAMM, L., MCKEOWN, G. G. and MORRISON SMITH, D. (1965) *J. Assoc. Offic. Agric. Chem.*, **48**, 892.
46. NICHOLAS, R. A. and FOX, J. B. (1973) *ibid.*, **56**, 922.
47. FIDDLER, R. N. and GENTILCORE, K. M. (1975) *ibid.*, **58**, 1069.
48. HAMILTON, J. E. (1976) *ibid.*, **59**, 284.
49. HARTLEY, A. M. and BLY, R. M. (1963) *Anal. Chem.*, **35**, 2094.
50. WESTCOTT, C. C. (1971) *Food Technol.*, **25**, 49.
51. PFEIFFER, S. L. and SMITH, J. (1975) *J. Assoc. Offic. Anal. Chem.*, **58**, 915.
52. LIEDTKE, M. A. and MELOAN, C. E. (1976) *J. Agric. Fd Chem.*, **24**, 410.
53. LOWE, R. H. and HAMILTON, J. L. (1967) *ibid.*, **15**, 359.
54. GARNER, G. B., BAUMSTARK, J. S., MUHRER, M. E. and PFANDER, W. H. (1956) *Anal. Chem.*, **28**, 1589.

55. McNamara, A. L., Meeker, G. R., Shaw, P. D. and Hageman, R. H. (1971) *J. Agric. Fd Chem.*, **19**, 229.
56. Lowe, R. H. and Gillespie, M. C. (1975) *ibid.*, **23**, 783.
57. Senn, D. R., Carr, P. W. and Klatt, L. N. (1976) *Anal. Chem.*, **48**, 954.
58. Klang, C.-H., Kuan, S. S. and Guilbault, G. G. (1978) *ibid.*, **50**, 1319.
59. Davenport, R. J. and Johnson, D. C. (1974) *ibid.*, **46**, 1971.
60. Bunton, N. G. and Tait, B. H. (1969) *J. Am. Wat. Wks Assoc.*, **61**, 357.
61. Horwitz, W. (Ed.) (1975) *Official Methods of Analysis*, 12th edn, Association of Official Analytical Chemists, Washington, DC, Section 20.029, p. 353.
62. *ibid.*, Section 20.033, p. 354.
63. *ibid.*, Section 20.030, p. 353.
64. Pearson, D. (1976) *The Chemical Analysis of Foods*, 7th edn, Churchill Livingstone, Edinburgh, London and New York, p. 39.
65. Horwitz, W. (Ed.) (1975) *Official Methods of Analysis*, 12th edn, Association of Official Analytical Chemists, Washington, DC, Section 20.042, p. 355.

PART 2
ORGANIC PRESERVATIVES

BENZOIC ACID AND THE BENZOATES

The water-soluble salts of benzoic acid permitted by UK regulations are those of sodium, potassium and calcium. Optimum activity is at pH 2·5–4·0, and thus the benzoates are useful preservatives in acid foods such as fruit juices, pickles, sauces and yoghurts, and since they are relatively weak acids it is evident that the undissociated form will predominate.

Commonly used extraction methods for the organic acids generally include steam distillation or solvent extraction of a salt-saturated and acidified food matrix. The general preference is for a combination of sodium chloride and phosphoric acid. It should be noted that certain small fruits contain benzoic acid as a natural constituent; amounts varying from 200 mg/l to 1400 mg/l have been reported in the juice expressed from cranberry and foxberry,[1,2] amounts up to 250 mg/l have also been reported in other fruits such as raspberries, plums and apricots.[3,4]

Qualitative Tests
As indicated above, most of the extraction procedures are based either on solvent extraction or steam distillation from salt/acid mixtures.[5,6] Tests utilising the chemical reactions of benzoic acid or UV (ultraviolet)

absorption spectra are carried out on an extract collected into sodium hydroxide and purified by reaction with a slight excess of permanganate at 60 °C. Excess of permanganate is removed with sulphur dioxide, and the presence of benzoic acid is demonstrated either by the modified Mohler test[7] or by dissolution of the crystallised material in ethanol and observation of the UV absorption spectrum.[8] The latter method is claimed to be applicable to the detection of benzoic, cinnamic, dehydroacetic, p-chlorobenzoic, salicylic and sorbic acids and the esters of p-hydroxybenzoic acid.

The Mohler test[9] (formation of dinitrobenzoic acid and reduction to diaminobenzoic acid) is subject to interference from the presence of salicylic acid and cinnamic acid. Test procedures for checking interference are based on the reaction of salicylic acid with ferric chloride and on an after-test which destroys the colour formed by benzoic acid in the Mohler reaction but not that formed by the other two acids.[7]

Although still widely used and officially recognised, the above test procedures are now being replaced by multi-detection screening methods based on chromatographic separations of the various organic acids. One such procedure has been elaborated by Tjan and Konter,[10] in which the preservatives are extracted with petroleum ether–ether (1:1), converted into sodium salts with sodium bicarbonate, and the acids separated from the esters of p-hydroxybenzoic acid by phase separation. The acid preservatives are spotted on silica gel G plates and developed in a mobile solvent containing ammonia; bromocresol purple is used as spray reagent. The esters are spotted on silica gel G, developed with benzene–ethyl acetate (4:1) and viewed under UV light (fluorescent indicator plate) or sprayed with bromocresol purple (plain plate).

An international collaborative study of a multi-detection TLC (thin-layer chromatography) procedure for detection of formic, propionic, trichloracetic, bromoacetic, dehydroacetic, salicylic, sorbic and benzoic acids and the esters of p-hydroxybenzoic acid has been carried out jointly between the AOAC and the Codex Alimentarius Committee on Methods of Analysis and Sampling. The method utilised solvent extraction procedures and two TLC systems; the first based on silica gel/cellulose with a petroleum ether–carbon tetrachloride–chloroform–formic acid–acetic acid mixture as developer, the second based on Polyamide 11 with a benzene–n–hexane–acetic acid–methylethylketone developer. Specific spray reagents were utilised to detect the individual preservatives, it is understood that some false positives were obtained by collaborators and that a further study and report is under way.[11]

Quantitative Methods

The official general methods recommended by the AOAC are based on titrimetry[12] and spectrophotometry.[13] Variations on extraction procedure are elaborated for different foodstuffs; in the first case the salted, acidified food preparations are extracted with successive amounts of chloroform, the chloroform is removed in a current of air and the residue dissolved in neutral alcohol and titrated with 0·05 N NaOH.

The spectrophotometric method applicable to tomato products, sauces, jams, jellies, soft drinks and fruit juices uses an ether extraction from an acidified, salt-saturated product; the acid is removed from the ether by treatment with ammonia, re-acidified and dissolved in ether. The ethereal solution is placed in a tightly stoppered cell and absorbance measurements are made at 272 nm (maximum) and 267·5 nm, 276·5 nm (minima). The method is applicable in the range 20–120 mg/l benzoic acid in the ethereal extract, but confirmatory evidence on absence of interferences may be necessary with certain products.

A quantitative TLC procedure utilises extraction by steam distillation into alkali and back-extraction into chloroform. TLC separation is achieved on kieselguhr/silica gel with n-hexane–acetic acid (96:4). The acid is located under UV radiation, the spot is removed by scraping and the acid is dissolved in ethanol. An absorbance measurement at 272 nm is used to obtain a quantitative assessment of the benzoic acid present.[14] Alternatively, reflectance spectrometry may be used to quantify the amounts present in the spots, limits of detection vary around 0·2 μg.[15] Inoue and co-workers[16] state that infrared detection at 1690 cm^{-1} and determination of benzoic acid in carbon tetrachloride is more advantageous than UV absorption and volumetric methods. It should be noted that UV absorption spectra are dependent on the type of solvent used; for benzoic acid the absorption maximum is at 272 nm, whereas in polar solvents such as methanol two maxima are observed at 228 nm and 272 nm. Roos and Versue[17] and Intonti et al.[18] used the two-maxima method for determination of benzoic acid in butter and margarine and in jams.

A spectrophotometric method for application to milk has been developed by Wahbi and co-workers.[19] This uses the first-derivative spectrophotometric curve obtained on milk diluted and clarified with citric acid, a filtered aliquot being further diluted with 0·5 N HCl. Readings are taken over the range 220–290 nm and at mean wavelengths chosen to suit the experimental spectra. The authors are able to obtain assays of benzoic acid, sorbic acid and propyl-p-hydroxybenzoate over concentration ranges of 2–10 mg/l, 1–5 mg/l and 2–10 mg/l, respectively. Recoveries obtained varied from 99·7% to 101·0% with standard deviations from 1·0 to 2·4. Direct observation techniques have also been applied to diluted soft drinks,[20] whilst Woidich et al.[21] have shown that an aliquot may be taken from the distillate obtained in the pycnometric determination of alcohol in wine for determination of benzoic acid and sorbic acid. Observations are made at 230 and 314 nm, and benzoic acid may be determined down to 2 mg/l. Zonneveld eliminated interferences by acid dichromate oxidation.[22] Interest in colorimetric methods is also maintained since simple colour tests are suitable for quality-control purposes. In control of margarine and edible oils, Kanematsu et al.[23] have used the oxidation of benzoic acid to salicylic acid with hydrogen peroxide as a means of using the sensitive colour reaction with ferric chloride; these workers claim that the method is less affected by impurities contained in the sample than the UV absorption or titration methods. A critical examination of the nitration reaction has been carried out by Hadorn and Doevelaar;[24] these workers showed that the products vary according to the conditions of the nitration reaction. This is important since the two colorimetric methods using nitration as the first stage depend on different derivatives; that due to Mohler[9] relies on 2,5-dinitrobenzoic acid, whilst that due to Spanyar[25] depends on the formation of 3,5-dinitrobenzoic acid; similar considerations also apply if diazotisation and coupling reactions are also to be used.

An automatic colorimetric method has been developed by Gend,[26] which relies on flash distillation from an acidified stream of liquid product, the benzoic acid being reacted with hydrogen peroxide in the presence of buffered cupric ion, catechol and sorbic acid. Colour development with 4-aminoantipyrine allows for determination of benzoic acid in the range 0–100 mg/l. Good precision is claimed.

Gas chromatographic methods have been used extensively to determine benzoic acid, early methods[27] were carried out using methyl esters prepared by reaction with diazomethane; silicone gum SE52 and PEG succinate were commonly used for separations of the mixed acid esters derived from steam-volatile acid preservatives. Chromatography of the free acids has been reported using SE-30 on Chromosorb W[28] for meat products, and Chromosorb W treated with hexamethyldisilazane and coated with Carbowax 20M[29] for analysis of oils. Trimethylsilyl derivatives have also been used for chromatography of the common acid preservatives.[30–32]

High-pressure liquid chromatography (HPLC) has been introduced for the determination of benzoic acid in soft drinks and other similar preparations, in which case the samples are injected directly onto the chromatographic column. Nelson[33] has used an anion-exchange column for the determination of benzoic acid, saccharin, p-hydroxybenzoates, xanthines and vanillin, whilst Smyly et al.[34] have used reversed-phase columns for determination of benzoic acid, saccharin and caffeine; recoveries better than 99 % for all three additives are claimed.

Collaborative Studies

The methods recognised by the AOAC have been subjected to collaborative study; the spectrophotometric procedure by Stanley,[35] the TLC procedure by Lewis and Schwartzman,[14] the titrimetric procedure by Sowbhagya et al.[36]

HYDROXYBENZOIC ACID AND DERIVATIVES

Hydroxybenzoate Esters

The three alkyl esters of p-hydroxybenzoic acid (parabens) have been allowed for use as food preservatives in the UK since 1962. They are stable against hydrolysis during autoclaving and resist saponification, so that they can be dissolved in 5 % sodium hydroxide. The anti-microbial activity is directly related to chain length but since solubility decreases with increasing chain length the lower esters are commonly used. The parabens have the advantage of a wider pH range of activity than benzoic acid.

Qualitative Analysis
Millons reagent and Deniges reagent have been used to detect the presence of the parabens and the free acid in foods. The free acid is not steam-volatile but the esters are partially so, hence the normal extraction procedure is by solvent from an acidified food. The solvent normally used is ether. Millons reagent gives a rose-red colour with the neutral ammonium salt of p-hydroxybenzoic acid, and hence it is necessary to hydrolyse the esters with alcoholic potash before carrying out the test.[37] Salicylic acid also gives a positive reaction, and prior testing with ferric chloride eliminates the possibility of a false positive.

Deniges reagent has been used as a spray reagent in the TLC separation of the free acid and the methyl and propyl esters.[38] Tammilehto and Buechi[39] claim that the reaction of p-hydroxybenzoic acid with aminoantipyrine is sufficiently specific for the detection of the free acid and the esters after saponification. The method of Tjan described in the section on benzoic acid has also been utilised to separate the parabens esters.

Reversed-phase TLC on silanised silica gel using ether- or ethylacetate-saturated borate at pH 11 has been used for separations of methyl, ethyl, propyl, butyl and benzyl esters.[40]

Quantitative Methods

A number of references have been made to papers with methods suitable for determination of mixed preservatives, and the following citations relate to methods described for application to parabens and the free acid.

Elution of esters from TLC plates with methanol or ethanol has been used prior to spectrophotometry for their determination[40-42] but gas chromatography appears to be a more popular technique. Lewis[43] used a 10% SE-30 column with helium carrier gas, and he recommended that further study was necessary before the method could be adopted as official. Donato[44] prepared the trimethylsilyl ethers of the methyl and propyl esters prior to gas chromatography (see also Larsson and Fuchs in benzoic acid). Kato[45] preferred trifluoroacetylation as a means of reducing retention times but found that the retention times of trimethylsilyl derivatives depended markedly on the polarity of columns. Daenens *et al.*[46] used a silica gel column to isolate the esters followed by elution with petroleum ether–ethyl ether and gas–liquid chromatography of the silyl derivatives. The method was claimed to be superior to that of Gossele[47] especially in its application to fatty foods; recoveries varying from 92% to 100% were obtained in application of the method to ketchup, salad mixes, salad dressings and pickles.

Polarography has been used by Tammilehto and Perala, who converted the esters into their 3-nitro derivatives[48] and 2,4-dinitrophenyl ethers.[49] A coefficient of variation of less than 4% was claimed for replicate determinations at the 0·5 mg level.

A continuous process analyser for measurement of *n*-heptyl-*para*-hydroxybenzoate in beer has been described by Saltzman and Wolf.[50] The instrument uses differential UV absorption measurement and wavelength shift with pH. The instrument is claimed to work in the range 0–15 mg/l with a precision of $\pm 0·2$ mg/l. A GC (gas chromatography) method[51] has also been described for application to beers by Courtney; this preservative has found use in brewing but it is not allowed in the UK.

Collaborative Studies

No records of collaborative studies other than the TLC procedure noted under benzoic acid have been recorded.

Para-chlorobenzoic Acid

This preservative appears to be effective in controlling mould and rope in bread; although not allowed in the UK it has been used in other countries. The acid responds to Mohlers test[52] and separations by TLC and GC procedures have after steam distillation or solvent extraction been noted by a number of workers (see especially Goodijn *et al.*,[27] Larsson and Fuchs,[30] and Adams,[11] noted in the section on benzoic acid).

Salicylic Acid
Salicylates were widely used as food preservatives but use is not permitted at the present in many countries including the UK. As a result, interest in development of quantitative methods has been limited. However, salicylic acid is generally included in the list of compounds studied in multiple-detection screening methods employing chromatographic procedures.

Qualitative Tests
The free acid may be extracted directly from acidified foods with ethyl ether; some pretreatment may be necessary to remove alcohol, proteins or fat.[53,54] The ferric chloride test may be confirmed by the use of Jorissens test yielding a red colour by reaction with copper sulphate in the presence of potassium nitrite and acetic acid. The procedure for detecting salicylic acid in the presence of benzoic acid has been described by Nicholls.[55] A direct method for application to fish products has been described by Hutshenreuter.[56] In this method, the protein is decomposed with phosphoric acid and the free salicylic acid is extracted with toluene, the colour reaction with ferric chloride being used.

Englis *et al.*[57] proposed a spectrophotometric method following ether extraction. Salicylic acid was also included in a number of studies using TLC methods noted in the section on benzoic acid.

Quantitative Methods
A solvent extraction method for application to vinegar has been described by Yamamoto *et al.*,[58] the acid being extracted into nitrobenzene in the presence of *tris*-(1,10-phenanthroline) iron (II) chelate, and absorbance is measured at 516 μm.

Official methods[53,59] utilise the ferric chloride reaction in a colorimetric procedure; the AOAC method includes a purification procedure using back-extraction into aqueous ammonia for application to the separated acid before the final determination. Salicylic acid has also been included in gas chromatographic procedures noted in the section on benzoic acid. The trimethylsilylation procedures described by Larsson and Fuchs[30] and Takemura[31] include provisions for salicylic acid.

Collaborative Study
No recent collaborative studies on quantitative methods have been recorded in relation to foodstuffs. Salicylic acid has been included in TLC screening procedures elaborated earlier.[10,11]

FATTY ACIDS AND DERIVATIVES

Dehydroacetic Acid
This acid is not permitted in the UK. However, there has been some interest in its use elsewhere in the prevention of mould growth in cheese, milk, beer and wine.

Qualitative Methods
Extraction with ether or chloroform is used to remove the dehydroacetic acid from foods. Purification by sublimation precedes reaction with Fehlings solution No. 1 to give a violet coloration;[60] alternatively, the chloroform extract of cheese is purified by extraction with sodium hydroxide, acidification and extraction into ether. The residue after removal of ether is dissolved in sodium hydroxide and reacted with salicylaldehyde in ethanol to yield a red or orange colour; the limit of detection is 10 mg/kg.[61] Titanium trichloride[62] has been reported to give an intense blue-violet colour with free acid. A chromotographic procedure for use with wines has been described.[63]

Quantitative Analysis
Spectrophotometric procedures for application in the analysis of cheese are described in the AOAC.[61] Residues in fruits have been determined by chloroform extraction and purification on silicic acid–glass fibre sheet using petroleum ether–ethyl ether–acetic acid as mobile solvent. The sodium salt is extracted and determined by UV absorbance measurement.[64] The reaction with *tris*-(1,10-phenanthroline) iron (II) chelate followed by extraction with nitrobenzene has also been used.[65,66] A fluorescent derivative is reported to be formed with boric acid, and the application to distillates from a range of samples is described.[67]

Collaborative Studies
A report of the collaborative study of the official method for cheese is given by Ramsey,[68] and recoveries varied from 59 to 103 %, average 89·7 % at the

level of 100 μg/kg. The method for fruits[5] has been studied collaboratively, with average recovery of 100 ± 5% over the range 5–80 μg/kg reported.

Haloacetic Acids and Haloacetates
The haloacetic acids have found use in wine and fruit products, but they are not permitted in the UK.

Qualitative Tests
Monochloroacetic acid is precipitated as a barium salt as hexagonal plates, and these may be observed under a polarising microscope. Two significant refractive indices may be measured $n\alpha(min) = 1\cdot582$ and $n\gamma(max) = 1\cdot611$. A confirmatory test is performed by reaction with anthranilic acid to form indigo.[69]

Mixed ether extraction from acidified wines is used prior to examination by TLC on silica gel G using benzene–dioxane–acetic acid, visualisation being achieved with diazobenzidine.[70] TLC on microcrystalline cellulose plates with dioxane–cyclohexane–acetic acid also provides a system for separation of chloro- and bromo-acetates. The plates are sprayed with ammonium carbonate and heated to 130 °C, ninhydrin spray then detects glycine as a reaction product.[71] Limits of detection of 1–2 mg/l are reported.

Quantitative Methods
A gas chromatographic method has been described for application to wines; the acids are extracted with ether from an acidified sample. GLC is carried out at 155 °C on a column packed with Chromosorb W loaded with 10% SE-30 using nitrogen as carrier gas.[72] The method is also suitable for the determination of benzoates, salicylates, chlorobenzoates and hydroxybenzoates.

Madrid[73] studied four methods for application to milk samples and concluded that a colorimetric method based on reaction with thiosalicylic acid to form thioindigo is preferred. Baluja et al.[74] reported that residues of phenoxyacetic acid herbicides are responsible for interferences in the thioindigo method due to the release of the acetic acid moiety which couples with thiosalicylate.

Collaborative Studies
No recent collaborative studies have been reported, and the AOAC methods date back to 1942–1949.

Propionic Acid
Propionic acid, together with its sodium and calcium salts, is permitted for use as an anti-mould and anti-rope agent in bread and flour confectionery. It is also used in other countries as an anti-mould agent in certain cheeses.

Qualitative Tests
Chromatographic methods are generally used to identify propionic acid in extracts prepared by steam distillation of the volatile acids. Young *et al.*[75] described a paper chromatographic method using acetone–butanol–ammonia as developing solvent; methyl red–bromothymol blue is used as a spray reagent. The method is suitable for separation of acetic, propionic, butyric and valeric acids and is included in the methods book of the AOAC.[76] TLC procedures have been described for application to bakery products[77] and to cheese.[78] Both methods identify the low-carbon-number volatile acids.

Quantitative Methods
Titrimetric methods have been applied after steam distillation and separation by silicic acid column chromatography.[79] Gas chromatography now appears to be the favoured technique for determination of the volatile fatty acids; a microdiffusion extraction technique with an average recovery of 98 % has been described by Karasz and Hallenbeck[80] whilst Graveland used ether extraction[81] followed by direct injection onto glass columns containing Chromosorb W coated with 5 % Carbowax 20M–terephthalic acid.

Cochrane[82] has described a system for application to the determination of C_2–C_{12} acids using prepared Chromasorb W loaded with neopentyl-glycol adipate, and the same author has also reviewed methods for the determination of C_2–C_6 acids.[83] A gas–solid chromatographic procedure using Porapack Q for application to cereal products has been described by Jones;[84] 100 % recoveries are claimed. A method for application to egg products has been described by Reagan *et al.*[85] which utilises the butyl esters for chromatography and includes succinic acid amongst the acids detected. An HPLC technique for application to bread with recoveries varying from 70 to 90 % has been described by Kanno *et al.*,[86] a limit of detection of 25 mg/kg being claimed.

Collaborative Studies
Young *et al.*[75] have reported on paper chromatographic detection and titrimetric analysis; Salwin[87] and Jones[84] have reported on gas chromatographic procedures.

Sorbic Acid

Sorbic acid is widely used as an agent which is more effective against yeasts and moulds than bacteria in flour confectionery, cheese, cider and perry, nut pastes, prunes, wines and wine vinegars.

Qualitative Tests

The acid may be separated by steam distillation from acidified foods; magnesium sulphate–sulphuric acid systems are recommended for general use[88,89] prior to UV absorption techniques or colorimetric tests. When measured in water in the pH range 2–4, sorbic acid shows a broad maximum at 261–264 nm; $E_{1cm}^{1\%}$ is normally accepted as 2260 under these conditions. A number of colorimetric reagents has been described, and these include phloroglucinol and thiobarbituric acid. The former has been utilised as a spray reagent in chromatographic systems,[90] whilst the latter is used directly on wine[91] or on steam distillates of other foods.[92,93] The reaction depends on prior formation of malondialdehyde by oxidation of sorbic acid with dichromate.

A number of TLC systems have been described for general application to mixed preservative systems. In most recent methods the spots have been located by examination under UV light.[94,95] Other composite methods are included in the section on benzoic acid. Methods using silica gel plates for application to bakery products,[96] wine[97] and fruit products[98] have been described.

Quantitative Analysis

Absorptiometric methods based on the use of thiobarbituric acid,[89,92,93] p-hydroxybenzaldehyde and other aromatic aldehydes[99] and more frequently on direct measurement in the UV have been used in the determination of sorbic acid in most foodstuffs. Iioka[100] has discussed possible interferences in the p-hydroxybenzaldehyde method arising from the presence of ginger and other spices in foodstuffs. The official method of the AOAC[89] depends on UV absorption at 250 nm in mixed ether solution; confirmation is dependent on oxidation with permanganate to destroy the sorbic acid.

A rapid extraction procedure with chloroform for application to dried prunes followed either by colorimetry or spectrophotometry has been described by Stafford,[101] and recoveries in the region of 98 % are claimed. Procedures for elimination of interferences in the analysis of wine[102,103] have also been described. Two automated procedures based on continuous distillation[104] and diffusion through a gas-permeable membrane[105] have been applied to liquid foods and to digests of solid samples. In both cases, the thiobarbituric acid colorimetric procedure is used.

Gas chromatographic methods for application to cheese,[106] fruit preparations[107] and feeds[108] have been described using various ester or silyl derivatives. Graveland[81] (see propionic acid) and Noda et al.[109] have described direct procedures on the extracted acids.

HPLC techniques have been used for mixed preservatives using silica gel with iso-octane–ether–propionic acid mobile phase.[110] McCalla et al.[111] have described a rapid method for application to wine. The sample is injected directly onto a column packed with the strong anion-exchange resin, Zipax SAX, the acid is extracted with 0·01 M borax and detected by absorbance at 254 nm. Levels as low as 10 mg/l can be detected with recoveries exceeding 90%. The method compares favourably with the AOAC procedure and an analysis is completed in 8 min.

Collaborative Tests
Collaborative studies and ruggedness tests[112] have been applied to the determination of sorbic acid in wine by the AOAC colorimetric and UV absorptiometric methods. Recoveries varied from 97 to 107%; coefficients of variation for the thiobarbituric method varied from 4·4 to 8·7% and from 2·0 to 6·4% for the UV absorption method. In further studies of the colorimetric procedure,[113] sources of erratic results have been identified, the major error being due to the imprecise instruction for quenching of the oxidation reaction. A revised wording for the method is proposed.

Wilamowski[114] has reported on a collaborative study for the determination of sorbic acid in cheese products using direct extraction with ether. Average recovery for the two cheese products used varied from 93·6 to 95·6% and from 90·7 to 101·9%. The AOAC method was found to have lower variability when applied to cheese.

OTHER PRESERVATIVES

Diethyl Pyrocarbonate (Diethyl Dicarbonate)
This preservative was developed for use in soft drinks, wines and beers to replace sulphur dioxide and benzoic acid; in these media most of the additive breaks down into ethanol and carbon dioxide, but in view of the formation of urethanes in wine it is not permitted in the UK. In the presence of ethanol, the pyrocarbonate undergoes solvolysis to yield diethyl carbonate.

Qualitative Tests
Diethyl dicarbonate is extracted with diethyl ether and reacted with 4-aminophenazone at 50 °C; dilution with water, addition of aqueous phenol and ammonia followed by potassium ferricyanide yields a red coloration.[115]

Quantitative Methods
The above colorimetric method may be used as an assay technique. Gas–liquid chromatography methods are generally used for determination of diethyl carbonate residues. Extracts are obtained with ether-, chloroform- or nitric-acid-treated carbon disulphide; typical column materials include 60–100 mesh Celite 545 loaded with 15% trimethylolpropantripelargonate, 60 mesh Firebrick C-22 loaded with 10–20% Carbowax 20 M, Chromosorb W loaded with 15% diethyleneglycol succinate polyester.[116-119] Studies on two new spectrophotometric methods for diethyl pyrocarbonate have been carried out by Berger,[120] but in this case the methods have been applied to biological fluids. The first reaction is based on a decolorisation of 5-thio-2-nitrobenzoate, the second on the inactivation of lactate dehydrogenase. The methods have been used to establish rate constants for the hydrolysis of diethyl pyrocarbonate.

Collaborative Studies
No collaborative studies have been recorded for the estimation of diethyl dicarbonate residues. Wunderliech[118] has studied the gas chromatographic method for estimation of diethyl carbonate in wine at levels from 0 to 10 mg/l. He concludes that the method is suitable for adoption as 'official'.

Formaldehyde
Formalin is a powerful preservative not permitted for use in foods; however, it is possible to find formaldehyde derived from wrapping materials in some foods or as a breakdown product of hexamethylene tetramine which is permitted in certain specified foods; traces derived from smoke constituents may also be found in smoked products.

Qualitative Tests
The tests may be directly applied to some products such as milk; solid and semi-solid samples are diluted with water and distilled after acidification with phosphoric acid.[121] Commonly used colorimetric tests include the chromotropic acid reaction[122] in which a purple colour is obtained and the

Hehner test which is based on the reaction between formaldehyde and tryptophan in the presence of sulphuric acid and ferric chloride. Milk may be used as the source of tryptophan. Alternative oxidising agents may be used. Fulton[123] claimed that the test is sensitive to 1 μg/ml if bromine is used. Iwami et al.[124] used the Hantzsch reaction in which formaldehyde reacts with acetyl acetone and ammonia to form a 3,5-diacetyl-1,4-dihydrolutidine condensation product both as a qualitative and quantitative method. The product is extracted into butanol to give a yellow layer. They claim that the method is directly applicable to fresh vegetables, mushrooms and dehydrated products.

Quantitative Methods
The acetyl acetone method is used in the official method for determination of formaldehyde in maple syrup; in this case a distillation procedure is used.[125] However, Karasz et al.[126] have proposed that the method can be used without the distillation step.

Modifications of the acetyl acetone method have been described by Czech[127] in which improved sensitivity is claimed. In this case simplex optimisation techniques were used to arrive at a limit of detection of 30 μg/kg with average relative standard deviation of about ±3%. A fluorimetric variation of this method has also been described.[128]

Alternative colorimetric procedures have been described based on chromotropic acid[129] and 7-amino-4-hydroxynaphthalene-2-sulphamic acid (J acid).[130] Simplex optimisation of both techniques has been carried out.[131,132] In the former case the limit of detection is 0·02 μg/ml, in the latter case an approximate improvement of × 5 over the original method is claimed. This is of the same order as the optimised chromotropic acid method. A rapid microdistillation method for application to feedstuffs has been described by Van Doren.[133] This also uses the chromotropic acid procedure. The Buchanan and Schryver method, reaction with phenylhydrazine and potassium ferricyanide, has recently been adopted as official in Germany.[134]

Collaborative Studies
Collaborative studies on the acetyl acetone method as applied to maple syrup have been published.[135,136] The latter followed work on the study of possible interferences in the method by gas chromatography[137] and mass spectrometry.[138] It was concluded that acetone and acetaldehyde are co-distilled with formaldehyde and that no interference exists.

Hexamethylene Tetramine (Hexamine)

Hexamine is allowed as a preservative in some overseas countries and in the UK as a preservative for Provolone cheese. It is determined as formaldehyde following distillation from an acidified solution.

Formic Acid

This acid is not allowed as a preservative in the UK. For analysis, steam distillation or ether extraction methods are used. Qualitative tests based on TLC procedures have been described.[139] Official quantitative methods[140,141] are based on distillation followed by gravimetric determination using mercuric chloride. It may also be determined by the chromotrophic acid method following percolation of an extract through a polyamide column.[142]

Biphenyl,2-Hydroxybiphenyl and Thiabendazole

Whilst biphenyl, 2-hydroxybiphenyl and thiabendazole (2-thiazol-4-yl benzimidazole) are used in the pre- and post-harvest treatment of certain agricultural crops and, as such, are more properly considered as pesticides (fungicides) rather than as food preservatives, at the present time they are included in the UK Preservatives Regulations.[143] In the EEC a similar situation exists, although there are proposals to transfer control from the Preservatives Directive[144] to an amended version of the Fruit and Vegetables Directive.[145]

As distinct from other food additives and contaminants, approved methods of analysis for residues of these compounds have been included as part of the legislative instrument. This review will also cover non-statutory methods which are, in some cases, superior with respect to recovery tests, detection limit or speed. Analytical methods for the detection and determination of fungicides have previously been reviewed by Baker and Hoodless.[146]

Qualitative Tests

Schedule 5, Part I, of the UK Regulations[143] describes a qualitative method to detect the presence of residues of biphenyl, 2-hydroxybiphenyl or the sodium salt, with limits of detection of 5 mg/kg and 1 mg/kg respectively. The compounds are extracted from the peel with dichloromethane and separated by TLC on silica gel plates. Spots are visualised under UV light and by colour tests. Similar TLC tests have been described for thiabendazole.[147,148]

Quantitative Methods

Schedule 4 of the Regulations[143] contains directions for the sampling of consignments, and the size of the sample to be taken for analysis to ensure that the sample taken is representative of the bulk package. Analytical methods for the quantitative determination of residues of biphenyl are described in Part II of Schedule 5 and for 2-hydroxybiphenyl in Part III of the same schedule.[143] These methods involve a preliminary distillation followed by solvent extraction, thin-layer chromatography and spectrophotometry. A critical evaluation of these methods has been published by Lord *et al.*[149] They made a number of suggestions for the improvement of the Statutory Method, particularly with respect to 2-hydroxybiphenyl where very low recoveries of added fungicide are obtained if the method is followed exactly as written. In addition, an alternative GLC procedure for both fungicides was proposed. Other workers have described methods based on both GLC[150-152] and spectrophotometry.[153-155] The application of HPLC techniques has been discussed by Farrow *et al.*[156] A number of other fungicides can also be determined using this method, although considerable clean-up to remove interfering co-extractives is still required.

Thiabendazole

Consignments of citrus fruits and bananas are usually examined for residues of thiabendazole by a surface-stripping technique using ethyl acetate. Maceration of the whole fruit has, in any case, been found to be a less efficient method of extraction.[147]

Determination of extracted residues of thiabendazole by a spectrofluorimetric method has been proposed[157] but benomyl and some citrus constituents may interfere. Norman *et al.*[158] introduced a TLC method to remove such interferences which could be used in the range 0·2–6 mg/kg of thiabendazole. They found no significant difference between the results obtained using the TLC and spectrofluorimetric methods. Alternatively, UV spectrophotometric methods have been developed by Mestres *et al.*[159] Thiabendazole exhibits a peak maximum around 300 nm with a characteristic and distinctive shoulder at 312 nm. Confirmatory tests using an enzymatic inhibition technique have been reported by Tjan and Burgess,[160] and Baker and Hoodless[161] designed a TLC/bioautographic screening test for benzimidazole compounds that can detect 2 mg/kg of thiabendazole. Alternatively, confirmation can be achieved using a GLC method with a sulphur-specific detector. However, this technique was found to be less accurate and lower in sensitivity.[162]

METHODS FOR DETECTION AND ESTIMATION OF MIXED PRESERVATIVES

In the foregoing text, a number of references have been made to multi-detection methods, many of which are based on the use of TLC and GLC, but there is now evidence that HPLC techniques are being used in this field. In addition to those methods noted in the text, a summary of which is presented in Table 1, a number of other methods are thought to be worth citation. Rajama and Makela[163] have developed a method which combines extraction and separation of benzoic acid, *para*-hydroxybenzoates and sorbic acid. Liquid foods are treated directly, solids are prepared as a slurry with water; the food is streaked across a strip of chromatographic paper, subsequently three parallel streaks of pH 6·5 buffer solution, sodium bicarbonate solution and sodium hydroxide solution are streaked at preset intervals along the paper. The food is then moistened with hydrochloric

TABLE 1

SUMMARY OF MULTI-DETECTION PROCEDURES

	References
Benzoic acid	7–10, 14–19, 21, 22, 27, 30, 32, 33, 41–43, 55, 81, 93, 94, 110, 163–171
Salicylic acid	7–11, 31, 42, 55, 63, 164, 166–170
Cinnamic acid	8, 9, 55
p-Chlorobenzoic acid	8, 168, 169
p-Hydroxybenzoic acid	8, 10, 38, 40, 163, 165, 167–170
p-Hydroxybenzoates	8, 11, 14, 15, 18, 19, 30–33, 38, 40–43, 72, 94, 110, 163–165, 168–170
Sorbic acid	8, 10, 11, 15–17, 19, 21, 22, 27, 30, 32, 41, 42, 70, 77, 81, 93, 94, 110, 163–165, 167–171
Formic acid	11, 87, 165, 170
Dihydroacetic acid	8, 11, 63, 65, 70, 164, 167–169
Chloroacetic acids	11, 71, 72, 169, 170
Bromoacetic acid	11, 70, 71, 168, 169
Propionic acid	11, 77, 81, 87, 170
Colorimetric/ absorptiometric methods	7–9, 16–19, 21, 22, 55, 65, 93, 163, 171
Paper chromatography	63, 163, 170
Thin-layer chromatography	10, 11, 15, 38, 40–43, 70, 71, 77, 94, 164, 165, 167, 168, 170
Gas–liquid chromatography	14, 27, 30–32, 43, 72, 81, 87, 169
High-pressure liquid chromatography	33, 110

acid, the paper is developed in diethyl ether. Ether-soluble organic acids are concentrated at the lower edge of the bicarbonate line and the *para*-hydroxybenzoates at the sodium hydroxide line; fat is transported to the solvent front. After 30 min development the paper is examined under UV to locate the preservatives; extraction with acidified ethanol removes benzoic acid and the *para*-hydroxybenzoates, extraction with water removes sorbic acid.

Quantitative assessment is by conventional techniques, and recoveries quoted are 75 % for parabens, 95 % for sorbic acid, 80 % for benzoic acid. A method for separation of parabens ester is also described.

Other general chromatographic systems have been described by Lee[164] using reversed-phase paper chromatography, Grune[165] TLC on silica gel, Fujiwara[166] ion exchange, Chiang[167] and Clement[168] TLC on polyamide, whilst Silbereisen[169] has described a two-column GLC system for 22 preservatives used in beer. The Nordic Committee on Food Analysis has also adopted general procedures using TLC and paper chromatography for 11 commonly used preservatives;[170] UV, quinine, bromophenol blue–methyl red and sulphanilic acid are used as visualising aids. Karasz *et al.*[171] have described a rapid screening method for determination of the common preservatives in ground beef; in such tests it is necessary to demonstrate the presence or total absence of preservatives and in this case a semiquantitative approach on a single sample extract is used. Sulphur dioxide is checked with *p*-rosaniline, ascorbic acid with 2,6-dichloroindophenol, benzoic acid and sorbic acid by UV absorbance measurements at 225 and 250 nm. If the presence of any preservative is indicated then quantitative assessment follows. These authors claim that the methods described are sensitive to 0·005 % sorbic acid and to 0·01 % for the other preservatives.

REFERENCES

1. VON GRIEBEL, C. (1910) Z. Untersuch Nahr. Genussm., **19**, 241.
2. CLAGUE, J. A. and FELLERS, C. R. (1934) Plant Physiol., **9**, 631.
3. NURSTEN, H. E. and WILLIAMS, A. A. (1967) Chem. Ind., 486.
4. ANON (1972) Annual Report of Tokyo Metropolitan Research Laboratory of Public Health, **24**, 249 (via Food Sci. Tech. Abstr. (1976), **8**, 9J, 1555).
5. PEARSON, D. (1976) The Chemical Analysis of Foods, 7th edn, Churchill Livingstone, Edinburgh and London.
6. MONIER-WILLIAMS, G. W. (1927) Analyst, **52**, 572.
7. ANON. (1975) Official Methods of Analysis of the Association of Official Analytical Chemists, 12th edn., 20.016–20.018.

8. TRIFIRO, E. (1960) *Ind. Conserve* (Parma), 279 (via *Biol Abstr.* (1961), **36**, 26573).
9. VON DER MOHLER, C. and JAKOB, F. (1910) *Z. Nahr., Genussm.*, **19**, 137.
10. TJAN, G. H. and KONTER, T. (1972) *J. Assoc. Offic. Anal. Chem.*, **55**, 1223.
11. ADAMS, W. S. (1978) *J. Assoc. Offic. Anal. Chem.*, **61**, 354.
12. AOAC, *ibid.*, 20.019–20.020.
13. AOAC, *ibid.*, 20.021–20.023.
14. LEWIS, M. H. and SCHWARTZMAN, G. (1967) *J. Assoc. Offic. Anal. Chem.*, **50**, 985.
15. DUDEN, R., FRICKER, A., CALVERLEY, R., PARK, K. H. and RIOS, V. M. (1973) *Z. Lebensm. U-Forsch.*, **151**, 23.
16. INOUE, T., KAWAMURA, T., KAMIJO, M. and ASAKURA, M. (1965) *Shokukin Eiseigaku Zasshi*, **6**, 154 (via *Chem. Abstr.* (1965), **63**, 15433d).
17. ROOSE, J. B. and VERSNEL, A. (1959) *Chem. Weekblad.*, **55**, 67.
18. INTONTI, R., RAMUSINO, R. C. and STACCHINI, A. (1960) *Boll. Lab. Chim., Prov.*, **11**, 147.
19. WAHBI, A. M., ABDINE, H. and BLAIH, S. M. (1977) *J. Assoc. Offic. Anal. Chem.*, **60**, 1175.
20. ENGLISH, E. (1959) *Analyst*, **84**, 465.
21. WOIDICH, H. and GNAUER, H. (1973) *Z. Lebensm. U-Forsch.*, **151**, 109.
22. ZONNEVELD, H. (1975) *J. Sci. Fd Agric.*, **26**, 879.
23. KANEMATSU, H., NINOMIYA, F., IMAMURA, M. and KAWAKITA, H. (1970) *J. Fd Hyg. Soc. Japan*, **11**, 463 (via *Food Sci. Technol. Abstr.*, (1972) **4**, 2N76).
24. HADORN, H. and DOEVELAAR, F. H. (1959) *Mitt. Lebensm. Hyg.*, **50**, 435.
25. SPANYAR, P., KEVEI, E. and KISZEL, M. (1958) *Z. Lebensm. U-Forsch.*, **107**, 118.
26. GEND, H. W. (1975) *Z. Lebensm. U-Forsch.*, **158**, 137.
27. GOODIJN, J. P., PRAAG, M. and HARDON, H. J. (1963) *Z. Lebensm. U-Forsch.*, **123**, 300.
28. CLARKE, E. G. C., HUMPHREYS, D. J. and STOILIS, E. (1972) *Analyst*, **97**, 433.
29. HILL, J. T. and HILL, I. D. (1964) *Anal. Chem.*, **36**, 2504.
30. LARSSON, B. and FUCHS, G. (1974) *Swed. J. Agric. Res.*, **4**, 109.
31. TAKEMURA, I. (1971) *Japan Analyst*, **20**, 61 (via *Anal. Abstr.* (1972), **22**, 4487).
32. FOJDEN, E., FRYER, M., and URRY, S. (1974) *J. Assoc. Publ. Anal.*, **12**, 93.
33. NELSON, J. J. (1973) *J. Chromatog. Sci.*, **11**, 28.
34. SMYLY, D. S., WOODWARD, B. B., CONRAD, E. C. (1976) *J. Assoc. Offic. Anal. Chem.*, **59**, 14.
35. STANLEY, R. L. (1960) *J. Assoc. Offic. Anal. Chem.*, **43**, 587; *ibid.* (1959) **42**, 486.
36. SOWBHAGYA, C. M., MAYURA, K., NAIR, K. E., SASTRY, L. V. L. and SIDDAPPA, G. S. (1963) *J. Assoc. Offic. Anal. Chem.*, **46**, 767; *ibid.* (1964) **47**, 68.
37. JOHNSON, H. W. (1946) *Analyst*, **71**, 77.
38. DICKES, G. J. (1965) *J. Assoc. Publ. Anal.*, **3**, 73.
39. TAMMILEHTO, S. and BUECHI, J. (1969) *Pharm. Acta Helv.*, **44**, 138.
40. RANGONE, R. and AMBROSIO, C. (1970) *J. Chromatogr.*, **50**, 436.
41. PINELLA, S. J., FALCO, A. D. and SCHWARTZMAN, G. (1966) *J. Assoc. Offic. Anal. Chem.*, **49**, 829.

42. LEMIESZEK-CHODOROWSKA, K. and SNYCERSKI, A. (1971) *Roczn. Panstu. Zak. Hig.*, **22**, 421. (via *Anal. Abstr.* (1972) **23**, 1937).
43. LEWIS, M. H. (1968) *J. Assoc. Offic. Anal. Chem.*, **57**, 876.
44. DONATO, S. J. (1965) *J. Pharm. Sci.*, **54**, 917.
45. KATO, S. (1968) *Eisei Shikenjo Hokoku*, **86**, 48 (via *Chem. Abstr.* (1970) **72**, 18283).
46. DAENENS, P. and LARUELLE, L. (1973) *J. Assoc. Offic. Anal. Chem.*, **56**, 1515.
47. GOSSELE, J. (1971) *J. Chromatogr.*, **63**, 429.
48. TAMMILEHTO, S. and PERALA, M. (1971) *Pharm. Acta Helv.* **46**, 351(via *Anal. Abstr.* (1972) **22**, 2667).
49. TAMMILEHTO, S. (1973) *Farmaseuttinen Aikak.*, **82**, 31 (via *Anal. Abstr.* (1974) **26**, 2267).
50. SALTZMAN, R. S. and WOLF, W. E. (1969) *Proc. Am. Soc. Brew. Chem.*, 201.
51. COURTNEY, J. M. (1968) *Proc. Am. Soc. Brew. Chem.*, 177.
52. WEISS, F. (1934) *Z. Untersuch. Lebensm.*, **67**, 84.
53. ANON. (1968) *Standard Methods Nordic Committee on Food Analysis*, No. 3.
54. ANON. (1975) *Official Methods of Analysis of the Association of Official Analytical Chemists*, 12th edn, 20.088.
55. NICHOLLS, J. R. (1928) *Analyst*, **53**, 19.
56. HUTSHENREUTER, R. (1956) *Z. Lebensm. U-Forsch.*, **104**, 14.
57. ENGLIS, D. T., BARNETT, B. B., SCHREIBER, R. A. and MILES, J. W. (1955) *J. Agric. Fd Chem.*, **3**, 964.
58. YAMAMOTO, Y., JUMAMURO, T., HAYASHI, Y. and TAKAHASHI, N. (1971) *Eisei Kagaku*, **17**, 109 (via *Chem. Abstr.* (1971) **75**, 62230g).
59. ANON. (1975) *Official Methods of Analysis of the Association of Official Analytical Chemists*, 12th edn, 20.091.
60. EECKHAUT, R. G. (1952) *Fermentatio*, 123.
61. ANON. (1975) *Official Methods of Analysis of the Association of Official Analytical Chemists*, 12th edn, 20.047–20.049.
62. SPERLICH, H. (1960) *Deut. Lebensm. Rundschau.* **56**, 70.
63. BASTIANUTTI, I. and ROMANI, B. (1960) *Boll. Lab. Chim. Exp.* (Bologna), **11**, 331 (via *Chem. Abstr.* (1961) **55**, 12761).
64. PIWOWAR, T. S. (1973) *J. Assoc. Offic. Anal. Chem.*, **56**, 1270.
65. HAYASHI, Y. (1971) *J. Sci. Hiroshima Univ.*, Ser. A-2, **35**, 147 (via *Chem. Abstr.* (1972) **77**, 18173h).
66. YAMAMOTO, Y., KAMMAMARU, T., HAYASHI, Y. and NOBORI, Y. (1967) *J. Pharm. Soc. Japan*, **87**, 1346 (via *Anal. Abstr.* (1969) **16**, 1551).
67. SHIBAZAKI, T. (1968) *Yakagaku Zasshi*, **88**, 1398 (via *Chem. Abstr.* (1969) **70**, 66853).
68. RAMSEY, L. L. (1953) *J. Assoc. Offic. Anal. Chem.*, **36**, 744.
69. ANON. (1975) *Official Methods of Analysis of the Association of Official Analytical Chemists*, 12th edn, 20.062–20.064.
70. COLONNA, C. (1974) *Riv. Viticoltura Enologia.* **27**, 280 (via *Fd Sci. Technol. Abstr.* (1975) **7**, 64873).
71. HALLER, H. E. (1971) *Deut. Lebensm. Rundschau*, **67**, 231.
72. REVUELTA, D., REVUELTA, G. and ARMISEN, F. (1975) *An. Quim.*, **71**, 179 (via *Anal. Abstr.* (1975) **29**, 4F28).
73. MADRID, A. (1972) *Rev. Espanola Lecheria*, **85**, 163 (via *Fd Sci. Technol. Abstr.* (1974) **6**, 3P304).

74. BALUJA, G., FRANCO, J. M. and MURADO, M. A. (1973) Arch. Environ. Contam. Toxicol., 1, 375.
75. YOUNG, J. A., SCHWARTZMAN, G. and MELTON, A. L. (1965) J. Assoc. Offic. Anal. Chem., 48, 622.
76. ANON. (1975) Official Methods of Analysis of the Association of Official Analytical Chemists, 12th edn., 14.091–14.093.
77. TJAN, G. H., JANSEN, J. TH. A. (1971) J. Assoc. Offic. Anal. Chem., 54, 1150.
78. ANTONINI, J. and ADDA, J. (1969) Ann. Techn. Agric., 18, 139 (via Food Sci. Tech. Abstr. (1970) 2, 5P62).
79. ANON. (1975) Official Methods of Analysis of the Association of Official Analytical Chemists, 12th edn, 14.086–14.090.
80. KARASZ, A. B. and HALLENBECK, W. (1972) J. Assoc. Offic. Anal. Chem., 55, 4.
81. GRAVELAND, A. (1972) J. Assoc. Offic. Anal. Chem., 55, 1024.
82. COCHRANE, G. C. (1973) Proc. Soc. Anal. Chem., 10, 212.
83. COCHRANE, G. C. (1975) J. Chromatogr. Sci., 13, 440.
84. JONES, F. B. (1973) J. Assoc. Offic. Anal. Chem., 56, 1415.
85. REAGAN, J. G., YORK, L. R. and DAWSON, L. E. (1971) J. Fd Sci., 36, 351.
86. KANNO, S., WADA, Y. and KAWANA, K. (1969) Shokuhin Eiseigaku Zasshi, 10, 371 (via Chem. Abstr. (1970) 73, 13193x).
87. SALWIN, H. (1965) J. Assoc. Offic. Anal. Chem., 48, 628.
88. CARR, W. and SMITH, G. A. (1964) J. Assoc. Publ. Anal. 2, 3.
89. ANON. (1975) Official Methods of Analysis of the Association of Official Analytical Chemists, 12th edn, 20.093–20.099.
90. ALLESSANDRO, A. (1967) Ind. Agr. (Florence), 5, 111.
91. LUECK, E. and REMMERT, K. H. (1974) Alimentaria, 11, 99.
92. DAVIDEK, J. and FRIMLOVA, E. (1962) Sb. Vysoke Skoly Chem-Technol. Praze., 6, 47 (via K. Sci. Fd Agric. (1964) 15, i–219).
93. GUTFINGER, T., ASHKENAZY, R. and LETAN, A. (1976) Analyst, 101, 49.
94. GOSSELE, J. A. W. and SREBRNIK-FRISZMAN, S. (1973) Mededelingen van de Faculteit Landbouwwetenschappen Rijksuniversiteit Gent, 39, 50 (via Fd Sci. Tech. Abstr. (1975) 7, 7T287).
95. BARBERO, L. (1973) Ind. Aliment., 12, 116 (via Chem. Abstr. (1974) 80, 35876).
96. VENTURINI, A. and ANGIULE, G. (1973) Boll. Lab. Chim. Provinciali, 24, 27 (via Chem. Abstr. (1973) 79, 90578c).
97. RIOS, V. M. (1972) Z. Lebensm. U-Forsch., 147, 331.
98. CHIANG, H. C. (1969) J. Chromatogr., 44, 201.
99. SALO, T. (1963) Suomen Kemistilehti, 36B, 1 (via Anal. Abstr. (1963) 10, 5396).
100. IIOKA, K. (1974) Eujo To Shokuryo, 27, 413 (via Chem. Abstr. (1975) 83, 7276s).
101. STAFFORD, A. E. (1976) J. Agric. Fd Chem., 24, 894.
102. MANDROU, B., ROUZ, E. and BRUN, S. (1975) Ann. Fals. Exp. Chim., 68, 29.
103. VAN BRONSWIJK, W. (1974) Aust. Wine Brew. Spirit Rev., 92, 40 (via Anal. Abstr. (1976) 30, IF20).
104. VAN GEND, H. W. (1973) Z. Lebensm. U-Forsch., 151, 81.
105. ROY, R. B., SAHN, M. and CONETTA, A. (1976) J. Fd Sci., 41, 372.
106. LA CROIX, D. E. and WONG, N. P. (1972) J. Assoc. Offic. Anal. Chem., 54, 361.

107. BONIFORTI, L., DI STEFANO, F. and VERCILLO, A. (1961) Boll. Lab. Chim. Provinciali, 12, 505 (via Anal. Abstr. (1962) 9, 3453).
108. RANFFT, K. and GERSTL, R. (1973) Z. Lebensm. U-Forsch., 151, 84.
109. NODA, K., KENJO, N. and TAKAHASHI, T. (1973) Shokuhin Eiseigaku Zasshi, 14, 253 (via Chem. Abstr. (1974) 80, 46588c).
110. WILDANGER, W. A. (1973) Chromatographia, 6, 381.
111. MCCALLA, M. A., MARK, F. G. and KIPP, W. H. (1977) J. Assoc. Offic. Anal. Chem., 60, 71.
112. CAPUTI, A. and SLINKARD, K. (1975) J. Assoc. Offic. Anal. Chem., 58, 133.
113. CAPUTI, A. and STAFFORD, P. A. (1977) J. Assoc. Offic. Anal. Chem., 60, 1044.
114. WILAMOWSKI, G. (1974) J. Assoc. Offic. Anal. Chem., 57, 675.
115. MONCELSI, E. (1970) Chim. Ind. (Milan), 52, 367 (via Anal. Abstr. (1971) 20, 2701).
116. TANI, T. and KANNO, S. (1970) Shokuhin Eiseigaku Zasshi, 11, 23 (via Chem. Abstr. (1970) 73, 13194y).
117. PRILLINGER, F. and HORWATITSCH, H. (1964) Mitt. hoh. Bundestehr u Vers Anst, Wein -Obst-u Gartenb. Rebe Wein, 14, 251 (via Anal. Abstr. (1966) 13, 990).
118. WUNDERLICH, H. (1972) J. Assoc. Offic. Anal. Chem., 55, 557.
119. PRILLINGER, F. (1964) Mitt. Klosterneuburg, Ser. A, Rebe Wein, 14, 29 (via Anal. Abstr. (1965) 12, 2507).
120. BERGER, S. L. (1975) Anal. Biochem., 67, 428.
121. ANON. (1975) Official Methods of Analysis of the Association of Official Analytical Chemists, 12th edn, 20.057.
122. EEGRIWE, E. (1937) Z. Anal. Chem., 110, 22.
123. FULTON, C. C. (1931) Ind. Eng. Chem. Anal. Ed., 3, 199.
124. IWAMI, K., YASUMOTO, K. and MITSUDA, H. (1974) Eiyo To Shokuryo, 27, 387 (via Chem. Abstr. (1975) 83, 7275r).
125. ANON. (1975) Official Methods of Analysis of the Association of Official Analytical Chemists, 12th edn, 31.175.
126. KARASZ, A. B., DECOCCO, F., MAXSTADT, J. J. and CURTHOYS, A. (1974) J. Assoc. Offic. Anal. Chem., 57, 541.
127. CZECH, F. P. (1973) J. Assoc. Offic. Anal. Chem., 56, 1496.
128. BELMAN, S. (1963) Anal. Chim. Acta, 29, 120.
129. BREMANIS, E. (1949) Z. Anal. Chem., 130, 44.
130. SAWICKI, E., HAUSER, R. T. and MCPHERSON, S. (1962) Anal. Chem., 34, 1460.
131. HOULE, M. J., LONG, D. E. and SMETTE, D. (1970) Anal. Lett., 3, 401.
132. CZECH, F. P. (1973) J. Assoc. Offic. Anal. Chem., 56, 1489.
133. VAN DOREN, P. H. (1976) J. Sci. Fd Agric., 27, 51.
134. MOEHLER, K. and DENBSKY, G. (1970) Z. Lebensm. U-Forsch. 142, 109.
135. UNDERWOOD, J. C. (1964) J. Assoc. Offic. Anal. Chem., 47, 548.
136. CALIXTE, H. (1973) J. Assoc. Offic. Anal. Chem., 56, 132.
137. UNDERWOOD, J. C. (1971) J. Assoc. Offic. Anal. Chem., 54, 30.
138. UNDERWOOD, J. C. (1972) J. Assoc. Offic. Anal. Chem., 55, 121.
139. ANON. (1967) Standard Methods, Nordic Committee on Food Analysis, No. 60.
140. ANON. (1966) Standard Methods, Nordic Committee on Food Analysis, No. 8.
141. ANON. (1975) Official Methods of Analysis of the Association of Official Analytical Chemists, 12th edn, 18.051.

142. LEHMANN, G. and LUTZ, I. (1970) Z. Lebensm. U-Forsch., 144, 318.
143. ANON. (1975) The Preservatives in Food Regulations, SI No. 1487, HMSO, London.
144. European Communities Secondary Legislation, Part 26, Directives 67/427 EEC and 71/160 EEC.
145. EEC Council Directive, 76/895 EEC, 23/11/1976, No. L 340, Office For Official Publications of the European Communities Luxembourg, 9/12/76, pp. 26–31.
146. BAKER, P. B. and HOODLESS, R. A. (1974) Pestic, Sci., 5, 465.
147. NORMAN, S. M., FOUSE, D. C. and CRAFT, C. C. (1972) J. Assoc. Offic. Anal. Chem., 55, 1239.
148. BAKER, P. B., FARROW, J. E. and HOODLESS, R. A. (1973) J. Chromatogr., 81, 174.
149. LORD, E., BUNTON, N. G. and CROSBY, N. T. (1978) J. Assoc. Publ. Anal., 16, 25.
150. TANAKA, A., NOSE, N., SUZUKI, T., HIROSE, A. and WATANABI, A. (1978) Analyst, 103, 851.
151. WESTÖÖ, G. and ANDERSON, A. (1975) Analyst, 100, 173.
152. BEERNAERT, H. (1973) J. Chromatogr., 77, 331.
153. GUNTHER, F. A., BLINN, R. C. and BARKLEY, J. H. (1963) Analyst, 88, 36.
154. RAJZMAN, A. (1963) Analyst, 88, 117.
155. WESTÖÖ, G. (1969) Analyst, 94, 406.
156. FARROW, J. E., HOODLESS, R. A., SARGENT, M. and SIDWELL, J. A. (1977) Analyst, 102, 752.
157. ANON. (1969) Pesticide Analytical Manual, Vol. II, Food and Drug Administration, Washington, DC, Section 120, 242.
158. NORMAN, S. M., FOUSE, D. C. and CRAFT, C. C. (1972) J. Agric. Fd Chem., 20, 1227.
159. MESTRES, R., CAMPO, M. and TOURTE, J. (1972) Ann. Fals. Exp. Chim., 65, 315.
160. TJAN, G. H. and BURGESS, L. I. (1973) J. Assoc. Offic. Anal. Chem., 56, 223.
161. BAKER, P. B. and HOODLESS, R. A. (1973) J. Chromatogr., 87, 585.
162. DE VOS, R. H. and BOSMA, M.P.M.M. (1970) Report No. R3199, Centraal Instituut voor Voedingsonderzoek TNO, Zeist, Netherlands.
163. RAJAMA, J. and MAKELA, P. (1973) J. Chromatogr., 76, 199.
164. LEE, S. C. (1969) Chemistry (Taipai), 94 (via Anal. Abstr. (1971) 20, 2082).
165. GRUNE, A. and NOBBE, V. (1967) Reichstoffe Arom. Korperpflegemittel, 17, 501 (via Anal. Abstr. (1969) 19, 1552).
166. FUJIWARA, M., MATSUMMURA, I. and FUJIWARA, K. (1971) Shokuhin Eiseigahu Zasshi, 12, 40 (via Chem. Abstr. (1971) 75, 34039h).
167. CHIANG, H. C. (1969) J. Chromatogr., 44, 203.
168. CLEMENT, J., VAN DIESSEL, L. and VAN KEYMEULEN, S. (1969) Fresnius Z. Anal. Chem., 248, 182 (via Anal. Abstr. (1970) 19, 5185).
169. SILBEREISEN, K. and WAGNER, B. (1970) Muenchen, Brau., 23, 57 (via Anal. Abstr. (1971) 20, 3400).
170. ANON. (1967) Standard Methods, Nordic Committee on Food Analysis, No. 60.
171. KARASZ, A. B., MAXSTADT, J. S., REHER, J. and DE COCCO, F. (1976) J. Assoc. Offic. Anal. Chem., 59, 766.

Chapter 5

MODE OF ACTION AND EFFECTIVE APPLICATION

ANTHONY J. SINSKEY

Massachusetts Institute of Technology, USA

INTRODUCTION

Food additives are most frequently understood to be chemical agents employed to extend the storage life of foods. Most additives are used mainly as preventative agents, i.e. to inhibit changes in flavour, nutritive value, odour, texture and other organoleptic properties of foods upon prolonged storage.

When a food processor develops and processes a food, the first priority is to control the microbial activity in the food from the viewpoint of food safety and quality. Any microbial or public health hazard has to be identified and properly controlled. Attention is next directed to control of chemical deteriorations, which can be broadly classified as either enzymatic or non-enzymatic. The chemical deteriorations of interest quite often consist of off-colour formations, i.e. browning reactions, lipid oxidation and pigment destruction. Finally, an effort is made to control the physical properties of the foods and food materials used as ingredients.

These broadly described issues, i.e. microbial, chemical and physical deterioration, are at first dealt with by the application of the basic principles of the science and technology of food preservation. Numerous fundamental and applied texts are now available to students and manufacturers, which describe the principles of food preservation by thermal processing, freezing, drying, concentration and irradiation. These texts include Karel *et al.*,[1] Nickerson and Sinskey,[2] Charm[3] and others. All of them concentrate on application of energy to inactivate and control public-health-significant or spoilage organisms.

111

However, other processing procedures also need to be employed because of a variety of constraints associated with certain types of foods and because of the lack of effectiveness of certain physical methods. One such procedure is the use of food additives, which when they are used mainly to retard microbial growth are commonly called food preservatives.

Although food scientists largely understand the application of physical processing procedures to foods, there are significant voids in our knowledge of food additives and food preservatives. These include:

(a) the prediction of how extrinsic and intrinsic parameters of a food system will control the effectiveness of a given food preservative;
(b) the mechanisms by which food preservatives inhibit or inactivate target micro-organisms; and
(c) the assessment of risk versus benefit to humans of long-term use of food preservatives.

The arguments most often put forward in defence of the use of food preservatives despite our admitted lack of knowledge about them include:

(a) their demonstrated technological necessity and effectiveness; and
(b) their safety in use, i.e. the fact that they are not carcinogenic or toxic at appropriate and realistic concentrations.

The health hazards of a food preservative are assessed by experimental procedures that follow a number of well defined steps. First of all, the food preservative must be identifiable by precise specification, so that toxicity data can be related to it. Recent developments in procedures and techniques of chemical analysis have resulted in marked improvements in analysis of food preservatives in foods.

In the next step, animal tests are designed and conducted, which reflect commercial usage and any eventual physical or chemical treatment of the additive-containing food. The effects of subsequent physical or chemical treatment processes on the behaviour of a food preservative should not be ignored in the evaluation programme. Observations must also be collected on humans exposed to the food preservative during its manufacture and technical use.

All available data are then assembled and extrapolated to determine the potential risk or health hazard to the consumer. The conventional toxicological testing schemes frequently vary in detail, but their overall features are standardised as described in Chap. 3.

Appreciation of the value of prokaryotic and eukaryotic cell systems has increased significantly in recent years. This is due in part to the development of a variety of short-term test procedures for the detection of mutations and other perturbations of cellular DNA. The value of these procedures lies in their sensitivities and low cost, as compared to conventional animal testing procedures, since the list of mutagenic chemicals which are also carcinogenic is continuously increasing. These short-term procedures can be used in conjunction with other more conventional tests in a tiered system or as part of a battery of tests for the detection of mutations as well as other types of genetic alterations. Malling[4] has reviewed the use of micro-organisms and cell cultures in testing for mutagenicity, and Nagao et al.[5] have summarised the current literature on mutation tests employing microbes for examining food and environmental samples.

Sinskey and Gomez[6] have recently (a) described briefly some of the prokaryotic and eukaryotic systems in current use; (b) indicated where and how such systems have been used to determine biological properties of food additives and food constituents; (c) indicated how such systems can also be used in conjunction with food processing research and development studies; and (d) speculated on possible research problems for the future.

Stimulated by safety considerations, more interest is now directed to improving the effectiveness of current food preservatives, and the desire to improve anti-microbial effectiveness has in turn stimulated interest in the mechanisms of food preservation. Where possible, recent research findings in this area will be summarised.

The purpose of this chapter is to discuss recent developments in food preservatives used primarily for the control of microbial proliferation. Consideration is given to mode-of-action studies as well as to location of use. For details regarding applications for specific products, the reader is referred to the articles by Kimble,[7] Chichester and Tanner[8] and Sinskey.[9]

SALTS

Sodium Chloride
Sodium chloride is one of the oldest known preservatives and is used in a variety of ways. For example, it can be used in low concentrations (2–4%), in combination with refrigerated storage or with acid, to inhibit the growth of spoilage organisms. Higher concentrations may also be employed, such as in the method of brining.

A variety of micro-organisms, i.e. halophilic, spoilage and pathogenic, are associated with various salted foods and seafoods. Lightly salted products, such as fish containing 1–4 % salt, may be readily spoiled by a variety of proteolytic and pathogenic micro-organisms such as *Clostridium botulinum*, *Clostridium perfringens*, *Staphylococcus aureus* and *Vibrio parahaemolyticus*.

The effectiveness of salt as a preservative is complicated by the intrinsic and extrinsic parameters of the food system. Whether or not salt is to be used as the main method of preservation, such treatment of foods must be carried out under controlled temperature conditions. If the temperature of the product immediately after salting is not below 15·6 °C, or preferably 4·4 °C, spoilage organisms including putrefactive bacteria may invade the tissues previous to penetration by the salt. This applies equally well to the growth of pathogenic bacteria such as *C. botulinum*.

Another factor involved in the inhibition of microbial growth by salting is available water, as micro-organisms grow only in aqueous solutions. The term 'water activity', A_w, has been coined to express the degree of availability of water in foods.[10] This term can be applied to all foods, with fresh foods having an A_w of about 0·99–0·96 at ambient temperatures. Low water activities, which limit the growth of micro-organisms in foods, may be achieved by the addition of salt or sugar, by the physical removal of water by drying, or by freezing.

According to Raoult's law,

$$\frac{P}{P_0} = \frac{N_2}{N_1 + N_2}$$

where P is the vapour pressure of the solvent (water in foods), P_0 is the vapour pressure of pure water, N_1 is the number of moles of solute present, and N_2 is the number of moles of solvent present. Furthermore,

$$\frac{P}{P_0} = A_w = \frac{\text{equilibrium relative humidity}}{100}$$

For example, a 1 M solution of a perfect solute in water would reduce the vapour pressure of the solution to 98·23 % of that of pure water. The A_w would be 0·9823.

Microbial water relationships have been reviewed by Brown[11] and water activity and food by Troller and Christian.[12] In general, moulds grow at lower water activities than yeasts, and yeasts grow at lower water activities than bacteria, but there are some exceptions. Bacteria usually require water activities between 0·99 and 0·96, but some non-halophiles will grow at water activities as low as 0·94. *Micrococcus halodenitrificans*, a xerophilic red halophilic organism, will grow at an A_w as low as 0·76, the A_w of a saturated sodium chloride solution.[11] *Halobacterium salinasium* requires approximately 3 M sodium chloride for growth, when this is the only solute. The physiology of these bacteria is dominated by an absolute requirement for sodium chloride.

The water relations of food-borne bacterial pathogens have been reviewed by Troller[13] and Troller and Christian.[12] Data indicate that in general food-borne bacterial pathogens can grow at water activity levels of 0·99 to 0·83.

The mechanisms by which micro-organisms adjust to alterations in water activity have been reviewed by Brown.[11] The intracellular physiology of extreme halophiles is dominated by massive accumulations of K^+ and Cl^- and by the effective exclusion of Na^+. With xerotolerant yeasts, polyhydric alcohols function as compatible solutes.

Some bacteria accumulate specific metabolites in response to water stress. Measures[14] has demonstrated an accumulation in bacteria of glutamate, γ-aminobutyrate, or proline, in response to increased salt concentration. A qualitative trend was observed in the type of amino acid accumulation, that is, glutamate predominated in the least salt-tolerant bacteria studied, whereas proline predominated in the most salt-tolerant. Thus, amino acids might well act as amino regulators under mild conditions and function as low-grade compatible solutes.

Thus, one result of adding salt or sugar to foods is to limit the growth of micro-organisms by lowering water activity. But there are other factors which play a role in limiting the growth of micro-organisms when salt or sugar is added. For instance, it has been shown[15] that when sodium chloride is used to adjust the A_w, *C. botulinum* types A and B will grow at water activities as low as 0·96, but that when glycerol is used instead of salt, these organisms will grow at water activities as low as 0·93. Thus, the inhibiting effect of sodium chloride is not due entirely to the lowering of water activity.

Nitrite

Nitrite is included in meat-curing mixtures primarily to develop and fix colour. Nitrites decompose to nitric oxide, which then reacts with haem pigments to form nitrosomyoglobin.[16]

In addition to fixation of colour, nitrite plays an important role in producing the characteristic flavour of cured meat. To produce the flavour of ham, which distinguishes it from salt pork, some 50 mg of nitrite per kg are thought to be necessary.

Nitrite alone or in combination with sodium chloride has important anti-microbial properties, as demonstrated in a number of recent publications.[16]

The use of nitrite as a curing agent provides protection against botulism.[17] It may also be important in inhibiting the growth of other food-poisoning and food-infection microbes, such as *C. perfringens*, *Bacillus cereus*, *S. aureus* and salmonellae. Nitrite is also effective against saprophytic bacterial spoilage.

The anti-bacterial effect demands greater additions of nitrite than is needed for the development of colour and flavour. According to Ingram,[16] more than 100 mg nitrite per kg is necessary to secure protection against botulism.

Mode of Action of Nitrite

The anti-microbial effectiveness of nitrite, a week acid salt, is dependent upon the pH of the food system, and to a large extent on the presence of the undissociated form HNO_2. The concentration of HNO_2 is related to pH, with a pK_a value of 3·4.

Nordin[18] has also reported that nitrite disappearance in meat or culture media is dependent upon acidity and temperature. The exponential decay relationship, as reported by Nordin,[18] is:

$\log_{10}$ (half-life of nitrite in hours)

$$= 0·65 - 0·025 \times \text{temp.(in } °C) + 35 \times pH$$

From this, it is seen that acidity and high temperature both shorten the half-life of nitrite in complex products. The use of filter-sterilised nitrite in the cold is warranted, in view of the effect of temperature on the disappearance of nitrite.

Ingram[16] has indicated that the pH resulting in a balance between formation and disappearance of HNO_2, which is optimal for anti-microbial action, may be around 5·5. Experiments have tended to confirm this result. For example, Tarr[19,20] showed that the preservative action of nitrite in fish is increased with acidification, and markedly so at levels below pH 6·0. As the pH is further lowered, anti-microbial effects begin to diminish around pH 5·5.[21] Shank *et al.*[22] also report an increase in anti-microbial effect with decreasing pH, with a maximum around pH 5·0.

It has long been known that anti-microbial properties of nitrites are complicated by a variety of factors. For instance, complex interactions of pH, NaCl, $NaNO_2$, $NaNO_3$, heat treatment (F_0 value), and incubation or storage temperature are known to be significant for the bacterial stability of cured meat products. Recent investigations have attempted to determine the extent and significance of particular interactions. A detailed study of the triple interaction of pH, NaCl and $NaNO_2$, and the influence of incubation temperature, has been quantitatively studied in a laboratory medium using *C. botulinum* types A, B, E and F.[23]

Although significant conclusions have been drawn from the above studies, including the fact that salt is the major inhibitory factor in cured meats and that heating is critical for development of a stable cured meat product, more knowledge of the inhibitory mechanisms of nitrite is desirable. Johnston *et al.*[24] listed four possible mechanisms for the inhibitory effect of nitrite on bacterial spores. They are:

(1) enhanced destruction of spores by heat;
(2) stimulation of spore germination during heating, followed by heat inactivation of germinated spores;
(3) inhibition of spore germination after heating; and
(4) production of more inhibitory substances from nitrite.

The first two mechanisms are not considered to be significant.[16] Inhibition of spore germination after heating has been described by Duncan and Foster.[25] Inhibition was dependent upon pH, and the undissociated NHO_2 again appears to be the active agent.

The fourth mechanism has attracted much interest recently. The historical developments are described by Ingram.[16] Briefly, it was observed that in some studies higher concentrations of nitrite than normally employed in industry were required for inhibition of spore outgrowth. Upon examination of the protocol, Perigo *et al.*[26] observed that these studies used filter-sterilised nitrite. It was therefore concluded that an inhibitory compound may be formed as a result of heating materials to which nitrite has been added.

This possibility was examined by two approaches; one concerned the development of inhibitory compounds during heating of bacterial culture media, and the other with meat systems.

A variety of studies have been conducted with heating of microbial culture media.[24,26–28] A general finding has been that, from nitrite and one or more components in the growth medium, a microbial growth inhibitor is formed, which is commonly called the 'Perigo inhibitor'. In a systematic study with *C. perfringens* as the test organism, Moran *et al.*[29] observed that only amino acids and mineral salts were involved in the production of the inhibitor, which was proven to be a compound formed from cysteine, ferrous sulphate and sodium nitrite. The inhibitor was compared to several known compounds, including S-nitrosocysteine, Roussin red salt and Roussin black salt. S-nitrosocysteine, although inhibitory, was not formed in sufficient quantities in the test system, whereas Roussin red salt was found to be unstable. Roussin black salt may have been formed. However, no single compound could be isolated, and the authors conclude that low levels of each compound may result in inhibition of microbial growth.

According to Ingram,[16] the Perigo inhibitor is not produced during the heat treatment of meat containing nitrite and cannot explain the role of nitrite in the stability of canned cured meats. Thus, further studies are required.

Nitrites have been shown to be involved in the formation of nitrosamines in cured meat products. The chemistry and health implications of nitrosamines have been discussed by Wogan.[30] Nitrosation of amines may take place in foods during processing or storage, and thus nitrosamines may be ingested in foods as such. In addition, it has been shown that nitrosation can take place in the strongly acidic conditions of the stomach. Therefore, ingestion of precursors can also lead to local formation of nitrosamines. These facts have stimulated considerable research and concern among public officials.

The key to the problem of nitrosamines in foods is nitrous acid. The genetic effects of nitrous acid have been reviewed by Zimmerman.[31] The chemical basis of nitrous acid mutagenesis is as follows. Nitrous acid is a medium–weak inorganic acid that is unstable in aqueous solution:

$$3HNO_2 \rightleftharpoons HNO_3 + 2NO + H_2O$$

Because of this instability, nitrous acid cannot be kept in free form. However, it can easily be generated by dissolving nitrite in a buffer below pH 5·5. The reaction sequence is the following:

$$NaNO_2 \rightleftharpoons Na^+ + NO_2^-$$
$$NaNO_2^- + H_2O \rightleftharpoons HNO_2 + OH^-$$

In the presence of surplus H^+ (buffer of low pH) the equilibrium is pushed to the right, thus increasing the concentration of the reactive free acid.

Undissociated nitrous acid can react with amino groups, and deamination of primary amines is rapidly achieved:

$$R—NH_2 + O{=}N—O—H \rightarrow R—N{=}N—OH + H_2O$$

Especially with aliphatic amines, the reaction proceeds further:

$$R—N{=}N—OH \rightarrow R—O—H + N_2$$

From secondary amines the more stable N-nitroso configuration is formed:

$$\underset{R_2}{\overset{R_1}{\diagdown}}N\!-\!H + H\!-\!O\!-\!N\!\!=\!\!O \rightleftharpoons \underset{R_2}{\overset{R_1}{\diagdown}}N\!-\!N\!\!=\!\!O + H_2O$$

The deamination reaction is the basis for the direct mutagenic action of nitrous acid, whereas the formation of N-nitroso compounds, such as N-alkylnitrosamines, is the basis for the indirect mutagenic activity via formation of mutagenic reaction products.[6,30,31]

Much research has been directed to determining the effectiveness of sodium nitrite in controlling outgrowth of *C. botulinum* and toxin formation from meat products. The effectiveness of sodium nitrite in delaying *C. botulinum* toxin production in cured meats has been demonstrated by Christiansen *et al.*,[32-34] and Hustad *et al.*,[35] who showed that its effectiveness increases with increasing nitrite concentration. Nitrite levels above 50 μg/g were found to delay botulinal toxin production.[35] Variations in the inhibitory effect of nitrite against *C. botulinum* toxin production was reported by Tompkin *et al.*[36] for meats ranging from turkey breast to pork hearts. The iron content of the various meats was shown to be an agent competing for available nitrite, which resulted in less nitrite for inhibition of spores.[36,37] This finding prompted the idea that the anti-microbial target for inhibition of spores was iron-containing cofactors or enzymes. It also explains why chelating agents such as ethylenediaminetetracetic acid increase anti-botulinal efficacy of nitrite. Isoascorbate and ascorbate enhance the anti-botulinal efficacy of nitrite in perishable canned cured pork if it is abused at the time of manufacture. This is due to the sequestering effect of isoascorbate or ascorbate on a cation, iron.[38] On the other hand, excessive levels of ascorbate have been shown to decrease the efficacy of nitrite in bacon.[38] The decreased inhibition occurs because isoascorbate and ascorbate cause more rapid depletion of residual nitrite. Thus, the use of isoascorbate has both positive and negative results in perishable canned cured meat products.

Recent investigations have shown that botulinal inhibition is possible even with lower nitrite levels (e.g. 40 μg/g) when sorbic acid or potassium sorbate is included in the formulation.[39-41] Studies by Sofos *et al.*[40] also indicated that a combination of nitrite (40 μg/g) and sorbic acid (0·2 %) was effective in delaying *C. botulinum* growth and toxin production in mechanically deboned chicken-meat frankfurter emulsions during temperature abuse. Similar studies have been recently reported with various meat and soy protein formulations.[41,42] Sorbic acid alone (0·2%) or in combination with nitrite (80 μg/g) retarded spore germination and outgrowth, and delayed toxin production.

The use of nitrite as a food preservative is still under debate from the viewpoint of risk versus benefit.

Sulphur-Dioxide-Generating Compounds

Sulphur dioxide and salts of sulphurous acid are used in foods to prevent enzymatic and non-enzymatic browning and the growth of undesirable micro-organisms. Sulphur dioxide also has anti-oxidant properties and is sometimes employed as a bleaching agent.[43]

Sulphur-dioxide-generating compounds include:

(1) sulphurous acid, H_2SO_3;
(2) the salts of sulphurous acid, including Na_2SO_3 (sodium sulphite), $NaHSO_3$ (sodium bisulphite) and K_2SO_3 (potassium sulphite);
(3) hydrosulphurous acid, $H_2S_2O_4$, the salt $Na_2S_2O_4$ (sodium hydrosulphite) being a strong reducing agent; and
(4) pyrosulphurous acid, $H_2S_2O_5$, and salt $Na_2S_2O_5$ (sodium pyrosulphite or metabisulphite).

The physical and chemical properties of sulphur-dioxide-generating compounds are described by Chichester and Tanner.[8]

Sulphur dioxide is highly soluble in water (11·3 g/100 ml at 20 °C), forming sulphurous acid (H_2SO_3), which can dissociate into the bisulphite (HSO_3^-) or the sulphite (SO_3^{2-}), depending upon pH:

$$SO_2 + H_2O \rightarrow H_2SO_3$$

$$H_2SO_3 \rightarrow H^+ + HSO_3^- \qquad pK_1 = 1·76$$

$$HSO_3^- \rightarrow H^+ + SO_3^{2-} \qquad pK_2 = 7·21$$

The specific concentrations of SO_3^{2-}, HSO_3^- and H_2SO_3 are dependent upon the pH of the solution, with HSO_3^- dominant between pH 2 and 7. Since it has been shown that sulphur dioxide is most effective as an anti-microbial agent in acid media, the anti-microbial agent is thought to be the undissociated sulphurous acid, which inhibits yeasts, moulds and bacteria. It has been reported that the undissociated acid is 1000 times more active than HSO_3^- for *E. coli*, 100–500 times for *Saccharomyces cerevisiae*, and 100 times for *Aspergillus niger*.[44]

The possible mechanisms by which sulphurous acid inhibits micro-organisms have been summarised by Lindsay[43] as follows: (*a*) reaction of bisulphite with acetaldehyde in the cell; (b) reduction of essential disulphide linkages in enzymes; and (c) formation of bisulphite addition compounds, which interfere with respiratory reactions involving nicotinamide dinucleotide. It should be pointed out, however, that most of the studies on the anti-microbial mechanisms of sulphite and sulphur dioxide were conducted a long time ago. Modern approaches of genetics and molecular biology have not been systematically applied.

Several studies have reported that sulphur dioxide and sulphites are metabolised to sulphate and then excreted in the urine without any obvious pathological results. For example, rats given oral doses of a 2% sodium metabisulphite solution eliminated 55% of the sulphur as sulphate within the first 4 hours.[43] In another study, groups of rats were fed sodium bisulphite in dosages from 0·01 to 2·0% of the diet for periods of 1 to 2 years. Rats fed 0·05% sodium bisulphite (307 ppm as SO_2) for 2 years showed no toxic symptoms. Sulphite in concentrations of 0·1% (615 ppm SO_2) or higher inhibited growth, probably through the destruction of thiamine in the diet.[43]

Treatment of foods with sulphites reduces thiamine content,[45] and it has been suggested that the ingestion of SO_2 in a beverage, for example, may cause a reduction in the level of thiamine in the rest of the diet.[8,9] Studies with rats have shown that the addition of SO_2 greatly reduces the urinary output of thiamine, especially when both are given together. Thus, sulphur dioxide and sulphite salts should not be used in foods which are substantial sources of thiamine. Levels above 500 ppm give a noticeably disagreeable flavour.[8,9]

Summers and Drake[46] reported that treatment of bacteriophage T4rII mutants with 0·9 M bisulphite for 4 hours resulted in mutations at guanine:cytosine (G:C) reversion sites specifically. Bisulphite was judged to be about as effective as nitrous acid applied at pH 5·3, but about 100 times weaker than nitrous acid applied at pH 3·7, in terms of mutations per mole per hour.[47]

In vitro studies[48] have reported on the conversion of cytosine to uracil in nucleic acid (1 M bisulphite). Bisulphite also adds to the 5:6 double bond of cytidine and uridine by means of an ionic reaction, forming 5,6-dehydrocytidine-6-sulphonate and 5,6-dehydrouridine-6-sulphonate, respectively.

Deamination of the cytidine derivative to the uridine derivative occurs readily, and under alkaline conditions the derivative is converted to uridine. In this manner, chemical modifications of transfer ribonucleic acid have been performed with bisulphite.[49]

A useful application of the fact that bisulphite deaminates cytosine to uracil is in localised mutagenesis. For example, Shortle and Nathaus[50] have developed a method for generating viral mutants with base substitutions in preselected regions of the viral genome. The clever aspect of the methodology is that only cytosine in single-strand nucleic acid is deaminated to uracil. Given some of the recent developments and procedures in recombinant DNA technology, this fact allows one to study mutations at very precise locations in DNA molecules.

All sources of bisulphite should be considered in the estimation of potential health risks. As pointed out by Fishbein,[47] these include not only bisulphite salts and SO_2 used as food preservatives, but bisulphite which is present in small amounts in polluted environments.

STERILISING AGENTS

Diethyl Carbonate
Diethyl carbonate, also called diethyl pyrocarbonate (DEPC), has been used, especially in Europe, for stabilising fruit juices, carbonated beverages and wines. Although its use was accepted for a number of years, it was banned by the Food and Drug Administration after Lofroth and Geguall[51] reported that ethyl carbamate was formed in wine thus treated.

Ethylene and Propylene Oxides
Although the epoxides, ethylene and propylene oxides, are used for gaseous sterilisation of spices, they are normally classified as anti-microbial agents.[8,9] Phillips[52] has reviewed the use of ethylene and propylene oxides as sterilising agents.

The desirability of the use of oxides for control of microbial levels lies not in the speed, simplicity or economy of treatment, but rather in the fact that, with certain types of foods, microbial levels are reduced with least damage to the food. Spices and nutmeats are prime examples. Also, since for effective use of the gases, the relative humidity of the products has to be low (30 % for both ethylene oxide and propylene oxide), their use is limited to low-moisture items. The chemical and physical factors affecting sterilisation by ethylene oxide are described by Ernst and Doyle.[53]

Stringent safety procedures are required for the use of these compounds. Ethylene oxide is volatile and flammable. Normally it is supplied as diluted mixtures. Propylene oxide is less reactive and has a narrower explosive range (2–22 %).

Concern regarding ethylene oxide sterilisation of foodstuffs has a history. Hawk and Mickelsen[54] demonstrated that rats did not grow when fed special purified diets treated with ethylene oxide. Reaction of ethylene oxide with various essential vitamins and amino acids was determined as the cause.

Gordon et al.[55] found that ethylene oxide alkylated cellulose in dry prunes, and glycols were also formed, as summarised by Phillips.[52] Concern over possible ethylene glycol formation led the FDA to ban the use of ethylene oxide treatment of foods for human consumption, while allowing the use of propylene oxide, since it, unlike ethylene glycol, is non-toxic. However, according to Phillips,[52] if salt is present in the product being treated, the more probable hydrolysis product is chlorohydrin rather than glycol, and both ethylene and propylene chlorohydrin are more toxic than ethylene glycol.

Hydrochloric Acid Vapour
In the continuing search for chemical agents that can be used for inactivation of micro-organisms on both glass and plastic surfaces, Tuynenburg et al.[56] screened a variety of compounds. A number of chemical sterilisers, such as 3-propanolide, ethylene oxide, ethylenimine, hydrogen peroxide, ozone, formic acid, methyl bromide, propylene oxide and 1-chloro-2,3-epoxypropane, were considered. None of the above was suitable because at low residual concentrations they are toxic and at working concentrations they form explosive mixtures or take too long to be effective at ambient temperature. Therefore, hydrochloric acid vapour was explored as an alternative approach for low-temperature sterilisation of plastic and glass containers.

The study of Tuynenburg et al.[56] showed that vapour evolving from hydrochloric acid can effectively inactivate spores of moulds, aerobic and anaerobic bacteria at low temperatures. The number of spores in a 300 ml glass bottle can be reduced within 5 minutes by at least a factor of 2500 at $20\,°C$ by 0.25 ml of $31\,\%$ (w/w) solution. These results are similar to those obtained with concentrated hydrogen peroxide solution at $24\,°C$.[57] The advantage of hydrochloric acid, however, is that it is not explosive and does not leave residues which may adversely affect the quality of a product, as may happen with hydrogen peroxide.[56]

Lelieveld and van Eijk[58] have investigated the influence of process parameters on biocidal activity as well as the use of dry hydrogen chloride. Process parameters studied were the influence of treatment time, temperature and partial hydrogen chloride and water vapour pressures on the sterilisation. Important factors affecting efficacy were size of dust particles, contact time and vapour pressure of HCl. The smaller the dust particles, the better the sterilisation, probably due to the lower degree of inhibition of diffusion of gas to the micro-organisms enclosed in the dust particles.

Hydrogen Peroxide

Hydrogen peroxide can protect food from spoilage by destroying or inhibiting the growth of micro-organisms. It is used as a bactericide in foods such as milk,[59] and as a bleaching and sterilising reagent in fish paste products and boiled noodles in Japan.[60] Toledo et al.[57] and Wallen and Walker[60] more recently have studied the sporicidal effects of hydrogen peroxide and the effects of media and media components on recovery of spores. Wallen and Walker's study indicates that the sporicidal action of hydrogen peroxide may not be as great as previously observed and that careful attention should be paid to the procedures used to evaluate the safety of food processes utilising hydrogen peroxide as a bactericide.

Carbon Dioxide

Another gaseous chemical that can be classified as a food preservative is carbon dioxide or carbonic acid. It is employed for the gas storage of fresh fruits, vegetables and animal products and for the preservation of fruit juices. It is also used in the bottling of soft drinks, beers and oleo and to replace oxygen in cans of dehydrated food products.

Yeast growth in juice can be inhibited by 1.5% (by weight) of carbon dioxide.[61] The effects of carbon dioxide on the shelf-life of a variety of meat products is still an active area of research and product development work.[62,63]

Ozone

As mentioned above, ozone has been studied for use as a sterilising agent. While most work has been directed to the use of ozone for disinfection of water and waste water, a limited number of studies have been conducted with food. For example, Haraguchi et al.[64] evaluated the preserving effect of ozone on fish and found that it destroyed a variety of moulds, yeasts and aerobic bacteria.

Yang and Chen[65] recently reported on the stability of ozone and its germicidal properties on poultry-meat micro-organisms in liquid phase. The stability of ozone in water depended on the water temperature, initial ozone concentration and length of holding time. In general, ozone was more stable in water at 2 °C than at 25 °C. As with other agents, the germicidal effects of ozone were influenced by contact time, temperature, pH value and presence of inorganic and organic materials in the disinfecting solution.

ORGANIC ACIDS

In addition to the inorganic acids, a variety of organic acids are used as food preservatives.[8,9,61] As with nitrite and sulphate, the greatest anti-microbial action is observed with the undissociated form of a given acid. Thus, from a practical viewpoint, the pK of the acid defines the pH range over which it may be expected to be effective as an anti-microbial. Solubility also plays an important role in determining the anti-microbial effectiveness of organic acids.

Many of the organic acids, such as acetic, lactic and citric acid, are used directly or indirectly to control the pH of a given food. It is normally impossible to lower the pH of a food product to the point where no micro-organisms will grow, since such products would have objectionable tastes. Thus, when used as preservatives the acids are usually employed in conjunction with other sublethal treatments, such as heat pasteurisation,

TABLE 1
COMMON FOOD ACIDS

Acid	pK	Maximal concentration present in some foods (mmol/kg)	Main spectrum of anti-microbial activity[b]	Food
Acetate (H, Na, K, di-)	4·76	500	m, b	Pickled food
Benzoate (H, Na)	4·2	8	m, y	Beverages
Parabens[a]				
Methyl	8·47	5		Beverages
Heptyl	8·47	?	broad	Beer
Propyl	8·47	0·6		Cream
Propionate (Na, Ca)	4·87	100	m, b	Cheese, bread
Sorbate (H, K, Na)	4·8	30	broad	Cheese, pies

[a] Esters of p-hydroxybenzoic acid.
[b] m = molds, y = yeast, b = spores of Bacillaceae.

refrigeration or salting. For example, acetic acid plays an important role in preservation of salted fish products, pickles and ketchup.

The acids most commonly used as food preservatives are listed in Table 1, along with the pK of each and an example of its use. For further details on established applications and properties, the articles by Sinskey[9] and Chichester and Tanner[8] may be consulted. This section will discuss more recent findings concerning the food acids.

Sorbic Acid

Many of the reports in the recent literature have been concerned with the use of sorbic acid as a food preservative. Scientists and technologists have recently recognised that the anti-microbial activities of sorbic acid and its salts are broader in scope than previously believed. This realisation has stimulated research on the use of sorbic acid in a variety of new products. Concurrent with the new uses has come the development of novel ways of applying sorbic acid, such as by spraying and dipping products into solution.

Sorbic acid and potassium sorbate preservatives are 'generally recognised as safe' (GRAS) for use in foods and are classified as such in the Code of Federal Regulations by the US Food and Drug Administration. They are used in a wide variety of food products, including processed cheese, wine, pie fillings and bakery products.

Kauk et al.[66] report on the influence of potassium sorbate on the microflora of butter at a variety of storage temperatures. Addition of 0·1 % potassium sorbate and of 2 % sodium chloride plus 0·1 % potassium sorbate resulted in inhibition of mould growth in all butter samples at the end of 4 weeks at −18 °C and 5 °C. However, the effect of potassium sorbate alone was less pronounced, irrespective of storage temperature.

Extensive studies have been reported on sorbic acid and potassium sorbate for use in cottage cheese.[67–69]

Robach[70] has reported on the extension of the shelf-life of fresh, whole broilers, using a potassium sorbate dip. Following a process wherein freshly chilled carcasses were dipped into a 5 % (w/v) solution of potassium sorbate for 30 seconds, the shelf-life of sorbate-treated birds was found to be 19 days at 3 °C, whereas for control birds it was 10 days. Robach and Ivey[71] have also reported that a 5 % potassium sorbate dip significantly reduced the total plate count of chicken breasts, as compared to counts from untreated breasts. The same authors have observed that a 5 % potassium sorbate dip markedly reduced growth rate of salmonellae inoculated onto the surface of the chicken breasts.

Robach and Hickey[72] have investigated the effectiveness of sorbic acid for inhibiting growth of three strains of *Vibrio parahaemolyticus* in two different seafood homogenates, crabmeat and flounder, under conditions favourable to the rapid growth of the organism (pH 6·2). When 0·1 % sorbic acid was incorporated into the homogenates, no increase in numbers of the three strains was observed in the crabmeat homogenate, and only slight increases in the flounder homogenate.

Another reason for the renewed interest in the use of sorbic acid as a preservative is that it may inhibit *in vitro* nitrosamine formation from amines and nitrite.[73] In addition, Ivey *et al.*[74,75] have shown that sorbic acid and potassium sorbate alone or in combination with low sodium nitrite can delay botulinal growth and toxin production in various meat products. Sofos *et al.*[39] investigated sodium nitrite effects on *C. botulinum* spore germination and total microbial growth in chicken frankfurter emulsions during temperature abuse. A much longer incubation period was necessary for the toxin to be formed in nitrite–sorbic acid treated samples as compared to controls or samples treated with nitrite or sorbic acid alone. Total microbial growth was not affected by the presence of nitrite, whereas sorbic acid appeared to inhibit it. In these studies sorbic acid was thought to inhibit *C. botulinum* spore germination. In the experiments mentioned in the section concerning nitrite, Sofos *et al.*[39] found botulinal spore germination and outgrowth to be significantly retarded by sorbic acid (0·2 %) and combinations of nitrite (80 μg/g) and sorbic acid (0·2 %) in all meat sources tested.

There are significant differences between studies as to where the challenge organisms are added to the products being evaluated. If inoculated during processing, different results are to be expected as compared to studies in which inoculation of test products occurs after processing. This point should not be overlooked by investigators comparing results from different laboratories.

Anti-microbial Mechanisms of Organic Food Acids

Although many of the chemical preservatives have been used for decades, their mode of action remains largely unexplained. Sinskey[9] has reviewed recent studies on the postulated mechanisms of inhibition of microbial growth by small organic molecules. These mechanisms can be broadly classified as follows:

(a) affecting cellular membrane integrity and function;
(b) affecting the genetic apparatus; or
(c) inhibiting specific enzymes.

The abundant indirect evidence for these mechanisms has been reviewed by Wyss.[76] More recent studies have shown interference with cellular membranes to be a primary mechanism of microbial inhibition. The papers by Sinskey,[9] Sheu and Freese,[77] Sheu et al.[78] and Sheu et al.[79] all point to membrane structure and function as the main targets. Important concepts to be derived from these papers are that (1) transport properties of membranes are altered by small organic molecules and (2) the cells wall can provide a protective barrier to the agent. Thus, chemicals or treatments which remove the cell wall, or lipopolysaccharide (LPS) layer surrounding the cell wall, improve the efficacy of an anti-microbial agent. For example, treatment of *Escherichia coli* with ethylenediamine tetraacetate (EDTA), which removes much of the LPS since it is bound ionically by salt bridges, renders the cell permeable to a variety of drugs. These findings have practical importance and probably explain the commonly observed phenomenon that in the presence of EDTA a number of food acids are more effective against a variety of gram-negative bacteria, including pseudomonas species.

Attempts have been made to correlate physical properties of food acids with their anti-microbial effectiveness. Sinskey[9] summarises some recent results with acids and alcohols. He shows that there is a good correlation between anti-microbial effectiveness against *Bacillus subtilis* and a plot of inhibitory concentrations of a variety of straight-chain alcohols, acids and diols versus a lipid/water partition coefficient. Freese[80] has summarised his own findings and those of others, and he comes to the conclusion that the inhibition of living cells, such as *B. subtilis*, by lipophilic acids is due to an inhibition of membrane transport of amino acids and other materials, which results in nutritional starvation of cells. The inhibition of transport results from the destruction of the proton motive force. To maintain a low or zero level of the proton gradient, the lipophilic acid must continuously shuttle protons into the cell and this requires the negative lipophilic ion to move back through the cell membrane. Thus, inhibitory potency of different lipophilic acids is determined by their lipid/water partition coefficients, their pK values and the ability of the molecules to delocalise the negative charge of the ions and thus increase membrane mobility. According to Freese,[80] the capacity of most lipophilic acids to delocalise the negative charge is weak enough for an equation to be derived relating the logarithm of inhibitory potency to the logarithm of the partition coefficient and a function describing the dependence on $(pH \cdot pK)$.

Other scientists have also reported on findings attempting to correlate structure and function of longer-chain fatty acids. For example, Kabara[81] makes the following points in reviewing his own findings and those of others:

(1) The most active chain length for saturated fatty acids is C_{12}; the most active monosaturated fatty acid is $C_{16:1}$; and $C_{18:2}$ is the most active polyunsaturated fatty acid.

(2) The *cis* form is more active than the *trans*.

(3) The position and number of double bonds is more important to long-chain (C_{12}) than to the shorter-chain fatty acids.

(4) Yeasts are affected by fatty acids with short chain lengths (C_{10}–C_{12}), which are slightly shorter than those most effective against gram-positive bacteria.

(5) Gram-negative organisms are affected by very-short-chain fatty acids.

(6) Fatty acids esterified to monohydric alcohol become inactive.

(7) Lauric acid esterified to a number of polyhydric alcohols becomes more active rather than less active. For example, lauric acid esterified with glycerol to form the mono-ester is very active. Di- and triglycerides are not active. This was the most important principle to be gleaned from the screening of some 300 lipophilic compounds.

(8) Acetylenic derivatives are more active against fungi as compared with ethylenic fatty acids.

These studies provide a renewed impetus for examining the anti-microbial properties of longer-chain fatty acids and mono-esters, which were previously discarded because of low anti-microbial activity. This is because the principal attribute of interest today is lack of toxicity rather than high anti-microbial activity. Both old and new principles of food science can be applied to maximise the effectiveness of these new sources of food preservatives.

ALCOHOLS

Alcohols, including polyhydric alcohols, are valuable food ingredients used in a wide range of food products. A discussion of the two-fold mechanism of inhibition of microbial growth by the polyhydric alcohols has been presented by Sinskey.[9] The first mechanism is a result of their ability to lower the water activity of foods or other suspensions, whereas the second is based on alterations in membrane structure and function.[9]

Ethanol can also be classified as a food preservative,[9] and there is currently a renewed interest in how ethanol inhibits microbial growth. This is due to the fact that ethanol produced from renewable resources for use in 'gasohol' requires for the most part rather economical fermentation processes. These processes produce ethanol in high concentrations, which may lead eventually to reduced recovery costs. Thus, studies of how ethanol inhibits microbial growth are of interest to a broad spectrum of scientists and engineers.

A number of physiological responses to ethanol have been characterised in different microbial systems.[82-86] A common feature of these findings is the primary role played by the cell membrane in regulation of ethanol tolerance. For example, E. coli cells respond to an ethanol challenge by altering the proportion of unsaturated fatty acids in the membrane.[82-84] This is due in part to the inhibition of saturated fatty acid synthesis. The viability of yeast cells in the presence of high ethanol concentrations has also been shown to depend upon membrane compositions.[85,86]

Yeast cell membranes artificially enriched with a polyunsaturated fatty acid (linoleic acid) result in yeasts that remain viable in ethanol to a greater extent than those enriched with oleic acid. The investigators concluded, based on this study and others, that a more fluid membrane results in an increased resistance to ethanol.

Clark and Beard[87] recently have employed mutants of E. coli with altered resistance to low-molecular-weight organic solvents to determine the mechanism of resistance or sensitivity to organic solvents. Solvent-resistant mutants showed a decrease in the ratio of phosphatidyl ethanolamine to the anionic phospholipids (phosphatidyl glycerol and cardyolipin) relative to the wild-type, whereas solvent-sensitive strains showed no increase.

EFFECTIVE APPLICATION OF FOOD PRESERVATIVES: THE ROLE OF ANTI-OXIDANTS, SEQUESTRANTS, INDIRECT ADDITIVES AND PHYSICAL TREATMENTS

The approaches used for effective microbial stabilisation of a food product were discussed in the introduction and involve application of food processing unit operations. With regard to chemical preservatives, several constraints need to be considered and some specifics have been mentioned. Thus, for an application to a specific food, one (a) defines the food and determines the legally accepted food preservatives that will present no organoleptic problems, (b) determines how the food preservatives shall be applied, e.g.

sprayed, dipped, mixed with ingredients or even encapsulated and (c) attempts to maximise the effectiveness of the food preservative.

Anti-microbial efficacy is improved by application and control of intrinsic and extrinsic parameters in a given food system. This is the only approach left to the food manufacturer, i.e. capitalising on the synergistic and additive effects of physical treatments or chemical additives. Unfortunately, the principles and data required for developing the stability maps that would be useful for a given food are not yet available. However, it is possible to list some recent scientific reports that may aid the food scientist in formulating a frontal attack on the maximisation of anti-microbial effectiveness in food preservatives.

Role of Physical Treatments in Improving Anti-Microbial Effectiveness
Shibasaki and Kato[88] and Takano et al.[89] have recently summarised studies on a variety of combined effects on the anti-bacterial activity of fatty acids and their esters against gram-negative bacteria. The impetus for their study was the fact that acetic, propionic and sorbic acids[90-92] and esters of p-hydroxybenzoic acid[93] are reportedly effective in combination with heating against *Salmonella* and yeast, although little data are available dealing with the middle- or long-chain fatty acids and their esters.

Lauric acid, monocaprin and monolaurin were found to be highly effective at low concentrations in combination with heating for inactivation of *E. coli* and *Pseudomonas aeruginosa*. Mild heating at 47 °C for short periods was found to decrease viable counts of both bacteria by factors of 10^{-3} to 10^{-6}. The interaction of freezing was also studied. When cells of *E. coli* or other bacteria were suspended in a nutrient broth containing Tween 20, Span 20, monolaurin or laurate and frozen to -20 °C, it was found that the viable counts were significantly reduced. Other investigators have also demonstrated the importance of 'cold shocking' on sensitivity of *E. coli* to fatty acids.[94] These studies clearly point to the importance of evaluating the longer-chain fatty acids and esters, many of which occur in foods, as anti-microbials in conjunction with physical treatments.

Role of Anti-Oxidants
The role of anti-oxidants as anti-microbials has also received attention. Trelase and Tompkin[95] demonstrated that butylated hydroxyanisole functions as an anti-microbial agent in food products and found it to be bactericidal at a concentration of 0·05 %, when tested for efficacy in controlling the normal spoilage flora of vacuum-packaged frankfurters. Butylated hydroxyanisole at 0·02 % was on the borderline, whereas 0·01 %

had little or no effect in controlling the spoilage flora of franks. Butylated hydroxytoluene was found to be ineffective. Turcotte and Saheb[96] also studied the anti-microbial activity of three anti-oxidants, butylated hydroxytoluene (BHT), butylated hydroxyanisole (BHA) and ethoxyquin (ETQ). BHT was most effective, and gram-positive bacteria were found to be more sensitive than gram-negative. Osmotic shock was observed to increase the sensitivity of *E. coli* to BHT. In addition, the increased anti-microbial effects of lauric acid or palmitoleic acid with BHT was reported in a continuation study.[97]

Sequestrants
Some sequestrants have also been shown to inhibit the growth of micro-organisms in foods.[9] EDTA has been shown to be an effective preservative for fish fillets.[98] As mentioned above, agents that alter the permeability of gram-negative cells tend to improve the effectiveness of a given anti-microbial. Besides EDTA, some investigators have shown that—in the presence of citric acid or polyphosphoric acid as well as EDTA—monolaurin, monocaprin, sucrose 'dicaprylate and sucrose monolaurate exhibited high bacterial activity against *E. coli* and other gram-negative bacteria.[88] It was concluded from experimental results that the transport of monolaurin into the cell membrane of *E. coli* was stimulated by the action of citric, polyphosphoric acid and EDTA.

As an example of the use of these approaches, it may be instructive to examine a recent patent by Nickerson and Darak,[99] which describes a composition and method for safely extending the storage life of refrigerated foods, such as shelled, hard-cooked eggs, cooked peeled shrimp, cooked and uncooked shrimp and cooked mushrooms. The approach used was to apply mixtures of preservatives with a buffering agent such as citric and/or phosphoric acids. Typical formulations proving effective were:

(1) a bacterial inhibitor selected from the class consisting of (*a*) approximately 0·08–0·20% methyl *para*-benzoic acid and (*b*) a mixture of approximately 0·075–0·15% sorbic acid with approximately 0·22–0·45% sodium propionate; and
(2) a buffering agent in amounts to maintain the pH of the solution in the range of approximately 4·5 to 5·5 and selected from the group consisting of (*a*) a mixture of citric acid and sodium or potassium hydroxide, (*b*) a mixture of sodium or potassium monohydrogen phosphate, dihydrogen phosphate or phosphoric acid.

Shelf-life of 6 months at 40 °F was reported for shelled hard-boiled eggs, whereas the shelf-life of the control was 5 weeks at 40 °F. Shelf-life of shrimp that were boiled, peeled and packed in the above described solutions at 40 °F was 4 months, whereas controls had a shelf-life of 4 weeks.

CONCLUSIONS

Anti-microbials will continue to be employed in a variety of foods. Intrinsic parameters such as pH, composition and water activity, in combination with environmental parameters such as temperature and gaseous storage atmosphere, will be employed to increase the effectiveness of the limited number of food preservatives that are currently available. In the opinion of this author, future research directions will be as follows.

(1) Development of more sophisticated stability maps for foods. By this is meant that the interactions between anti-microbial additives, environmental conditions and other factors should be described in terms of iso-surface responses to allow for prediction of shelf-life.

(2) Investigations into how organic molecules interact with the cell membrane. An important tool in this regard will be the use of high-resolution ^{31}P and ^{13}C nuclear magnetic resonance studies of whole cells *in vivo*. Recent papers by Ugurbil[100,101] among others point the way toward a most exciting and informative way of examining how anti-microbials interact with the cell membrane.

(3) Other naturally occurring food preservatives should be sought, in addition to the longer chain fatty acids.

(4) New delivery systems should be explored, including spraying, encapsulation and sustained release of the appropriate preservative.

REFERENCES

1. KAREL, M., FENNEMA, O. R. and LUND, D. B. (1975) In: *Principles of Food Science*, O. R. Fennema (Ed.), Part II, *Physical Principles of Food Preservation*, Marcel Dekker, New York.
2. NICKERSON, J. T. R. and SINSKEY, A. J. (1977) *Microbiology of Foods and Food Processing*, American Elsevier Publ. Co., New York.
3. CHARM, S. F. (1971) *Food Engineering*, 2nd edn, Avi Publ. Co., Westport, CT.
4. MALLING, H. V. (1978) In: *Handbook of Teratology*, Vol. 4, J. G. Wilson and F. C. Fraser (Eds.), Plenum Publishing Co., New York, pp. 35–69.
5. NAGAO, M., SUGIMURA, K. and MATSUSHIMA, T. (1978) *Ann. Rev. Genetics*, **12**, 117–59.

6. SINSKEY, A. J. and GOMEZ, R. F. (1980) In: *Food and Health: Science and Technology*, G. G. Birch and K. J. Parker (Eds.), Applied Science Publishers, London.
7. KIMBLE, C. E. (1977) In: *Disinfection, Sterilization and Preservation*, 2nd edn, S. S. Block (Ed.), Lea & Febiger, pp. 834–58.
8. CHICHESTER, D. F. and TANNER, F. W. (1968) In: *Handbook of Food Additives*, T. E. Furia (Ed.), The Chemical Rubber Co., Cleveland, OH, p. 137.
9. SINSKEY, A. J. (1979) In: *Nutritional and Safety Aspects of Food Processing*, S. R. Tannenbaum (Ed.), Marcel Dekker, New York, p. 369.
10. SCOTT, W. J. (1957) *Adv. Fd Res.*, 7, 83.
11. BROWN, A. D. (1976) *Bacteriol. Rev.*, 40, 803.
12. TROLLER, J. A. and CHRISTIAN, J. H. B. (1978) *Water Activity and Food*, Academic Press, New York.
13. TROLLER, J. A. (1973) *J. Milk Fd Technol.*, 36, 276.
14. MEASURES, J. S. (1975) *Nature* (London), 257, 398.
15. BAIRD-PARKER, A. C. and FREAME, B. (1967) *J. Appl. Bacteriol.*, 30, 420.
16. INGRAM, M. (1973) In: *Proc. Int. Symp. Nitrite Meat Prod.*, Zeist, p. 63.
17. CHRISTIANSEN, L. N., JOHNSTON, R. W., KAUTTER, D. A., HOWARD, J. W. and AUNAN, W. J. (1973) *Appl. Microbiol.*, 25(3), 357.
18. NORDIN, H. R. (1969) *Can. Inst. Fd Technol. J.*, 2(2), 79.
19. TARR, H. L. A. (1941 *Nature* (London), 147, 417.
20. TARR, H. L. A. (1941) *J. Fish Res. Brd Can.*, 5, 265.
21. HENRY, M. L., JOUBERT, L. and GORET, P. (1954) *Seanc-Soc. C-R Biol.*, 148, 819.
22. SHANK, J. L., SILLIKER, J. H. and HARPER, R. H. (1962) *Appl. Microbiol.*, 10, 185.
23. ROBERTS, T. A., JARVIS, B. and RHODES, A. C. (1976) *J. Fd Technol.*, 11, 25.
24. JOHNSTON, M. A., PIVNICK, H. and SAMSON, J. M. (1969) *Can. Inst. Fd Technol.*, 2, 52.
25. DUNCAN, C. L. and FOSTER, E. M. (1968) *Appl. Microbiol.*, 16, 401.
26. PERIGO, J. A., WHITING, E. and BASHFORD, T. E. (1967) *J. Fd Technol.*, 2, 377.
27. ROBERTS, T. A. and INGRAM, M. (1966) *J. Fd Technol.*, 1, 147.
28. RIHA, W. E. and SOLBERG, M. (1975) *J. Fd Sci.*, 40, 443.
29. MORAN, D. M., TANNENBAUM, S. R. and ARCHER, M. C. (1975) *Appl. Microbiol.*, 30, 838.
30. WOGAN, G. (1979) In: *Nutritional and Safety Aspects of Food Processing*, S. R. Tannenbaum (Ed.), Marcel Dekker, New York, p. 265.
31. ZIMMERMAN, F. K. (1977) *Mutation Res.*, 39, 127.
32. CHRISTIANSEN, L. N., JOHNSTON, R. W., KAUTTER, D. A., HOWARD, J. W. and AUNAN, W. J. (1973) *Appl. Microbiol.*, 25, 357.
33. CHRISTIANSEN, L. N., TOMPKIN, R. B., SHAPARIS, A. B., KEUPER, T. U., JOHNSTON, R. W., KAUTTER, D. A. and KOLARI, O. J. (1974) *Appl. Microbiol.*, 27, 733.
34. CHRISTIANSEN, L. N., TOMPKIN, R. B. and SHAPARIS, A. B. (1978) *J. Fd Prot.*, 41, 354.
35. HUSTAD, G. O., CERVENY, J. G., TRENK, H., DEIBEL, R. H., KAUTTER, D. A., FAZIO, T., JOHNSTON, R. W. and KOLARI, O. E. (1973) *Appl. Microbiol.*, 26, 22.

36. TOMPKIN, R. B., CHRISTIANSEN, L. N. and SHAPARIS, A. B. (1978) *Appl. Environ. Microbiol.*, **35**, 886.
37. TOMPKIN, R. B., CHRISTIANSEN, L. N. and SHAPARIS, A. B. (1978) *Appl. Environ. Microbiol.*, **35**, 863.
38. TOMPKIN, R. B., CHRISTIANSEN, L. N. and SHAPARIS, A. B. (1979) *Appl. Environ. Microbiol.*, **37**, 351.
39. SOFOS, J. N., BUSTA, F. F., BHOTHIPAKSA, K. and ALLEN, C. E. (1979) *J. Fd Sci.*, **44**, 668.
40. SOFOS, J. N., BUSTA, F. F. and ALLEN, C. E. (1979) *Appl. Environ. Microbiol.*, **37**, 1103.
41. SOFOS, J. N., BUSTA, F. F. and ALLEN, C. E. (1979) *J. Fd Sci.*, **44**, 1267.
42. SOFOS, J. N., BUSTA, F. F. and ALLEN, C. E. (1979) *J. Fd Sci.*, **44**, 1662.
43. LINDSAY, R. C. (1976) In: *Principles of Food Science*, O. R. Fennema (Ed.), Marcel Dekker, New York, p. 465.
44. REHM, H. J. and WITTMAN, H. (1962) *Z. Lebensm. Untersuch-Forsch.*, **118**, 413.
45. ARCHER, M. C. and TANNENBAUM, S. R. (1979) In: *Nutritional and Safety Aspects of Food Processing*, S. R. Tannenbaum (Ed.), Marcel Dekker, New York, p. 47.
46. SUMMERS, G. A. and DRAKE, J. W. (1971) *Genetics*, **68**, 603.
47. FISHBEIN, L. (1976) In: *Chemical Mutagens: Principles and Methods for their Detection*, A. Hollander (Ed.), Plenum, New York, p. 219.
48. SHAPIRO, R., SERVIS, R. E. and WELCHER, M. (1970) *J. Am. Chem. Soc.*, **92**, 422.
49. SINGHAL, R. P. (1971) *J. Biol. Chem.*, **246**, 5848.
50. SHORTLE, D. and NATHAUS, D. (1978) *Proc. Natl Acad. Sci. USA*, **75**, 2170.
51. LOFROTH, G. and GEGUALL, T. (1971) *Science*, **174**, 1248.
52. PHILLIPS, C. R. (1977) In: *Disinfection, Sterilization and Preservation*, S. S. Block, (Ed.), Lea & Febiger, Philadelphia, p. 592.
53. ERNST, R. R. and DOYLE, J. E. (1968) *Biotechnol. & Bioeng.*, **10**, 1.
54. HAWK, E. A. and MICKELSEN, D. (1955) *Science*, **121**, 442.
55. GORDON, H. T., THORNBURG, W. W. and WEREM, L. N. (1959) *Agr. Fd Chem.* **7**, 196.
56. TUYNENBURG MUYS, G., VAN RHEE, R. and LELIEVELD, H. L. M. (1978) *J. Appl. Bacteriol.*, **45**, 213.
57. TOLEDO, R. T., ESCHER, F. and AYRES, J. A. (1973) *Appl. Microbiol.*, **26**, 592.
58. LELIEVELD, H. L. M. and VAN EIJK, H. M. J. (1979) *J. Appl. Bacteriol.*, **47**, 121.
59. Food and Drug Administration, Dept HEW (1962) *Federal Register*, **27**, 3005.
60. WALLEN, S. E. and WALKER, H. W. (1979) *J. Fd Sci.*, **44**, 560.
61. VON SCHELHORN, M. (1951) *Adv. Food Res.*, **3**, 429.
62. BAILEY, J. S., REAGAN, J. O., CARPENTER, J. A., SCHULER, G. A. and THOMPSON, J. E. (1979) *J. Fd Prot.*, **42**, 218.
63. ROTH, L. A. and CLARK, D. S. (1975) *Can. J. Microbiol.*, **21**, 629.
64. HARAGUCHI, T., SIMIOU, U. and AISO, K. (1969) *Bull. Japan Soc. Sci. Fish*, **9**, 915.
65. YANG, P. P. W. and CHEN, T. E. (1979) *J. Fd Sci.*, **44**, 501.
66. KAUK, A., SINGH, J. and KUILA, R. (1979) *J. Fd Prot.* **42**, 656.

67. KRISTOFFERSEN, T. and CHAKABORTY, R. K. (1964) *J. Dairy Sci.*, **47**, 931.
68. HUANG, E. and ARMSTRONG, J. G. (1970) *Can. Inst. Fd Tech.*, **3**, 157.
69. BRADLEY, R. L., HARMON, L. G. and STINE, C. M. (1962) *J. Milk Fd Tech.*, **25**, 318.
70. ROBACH, M. C. (1979) *J. Fd Prot.*, **42**, 855.
71. ROBACH, M. C. and IVEY, F. J. (1978) *J. Fd Prot.*, **41**, 284.
72. ROBACH, M. C. and HICKEY, C. S. (1978) *J. Fd Prot.*, **41**, 699.
73. TANAKA, K., CHANG, K. C., HAYATSU, H. and KARA, T. (1978) *Fd Cosmet. Toxicol.*, **16**, 209.
74. IVEY, J. F., SHAVER, K. J., CHRISTIANSEN, L. N. and TOMPKIN, R. B. (1978) *J. Fd Prot.*, **41**, 621.
75. IVEY, J. F. and ROBACH, M. C. (1978) *J. Fd Sci.*, **43**, 1782.
76. WYSS, O. (1948) *Adv. Fd Res.*, **1**, 373.
77. SHEU, C. W. and FREESE, E. (1973) *J. Bacteriol.*, **115**, 869.
78. SHEU, C. W., KONINGS, W. N. and FREESE, E. (1972) *J. Bacteriol.*, **111**, 525.
79. SHEU, C. W., SALOMON, D., SIMMONS, J. L., SREEVALSON, T. and FREESE, E. (1957) *Antimicrob. Agents Chemother.*, **7**, 349.
80. FREESE, E. (1978) In: *The Pharmacological Effect of Lipids*, J. J. Kabara (Ed.), The American Oil Chemists Society, Champaign, IL, p. 123.
81. KABARA, J. J. (1978) In: *The Pharmacological Effect of Lipids*, J. J. Kabara (Ed.), The American Oil Chemists Society, Champaign, IL, p. 1.
82. INGRAM, L. O. (1976) *J. Bacteriol.*, **125**, 670.
83. INGRAM, L. O. (1977) *Appl. Environ. Microbiol.*, **33**, 1233.
84. BUTTKE, T. M. and INGRAM, L. O. (1978) *Biochemistry*, **17**, 637.
85. THOMAS, D. S., HOSSACK, J. A. and ROSE, A. H. (1978) *Arch. Microbiol.*, **117**, 239.
86. THOMAS, D. S. and ROSE, A. H. (1979) *Arch. Microbiol.*, **122**, 49.
87. CLARK, D. P. and BEARD, J. P. (1979) *J. Gen. Microbiol.*, **113**, 267.
88. SHIBASAKI, I. and KATO, N. (1978) In: *The Pharmacological Effect of Lipids*, J. J. Kabara (Ed.), The American Oil Chemists Society, Champaign, IL, p. 15.
89. TAKANO, M., SIMBOL, A. B., YASIN, M. and SHIBASAKI, I. (1979) *J. Fd Sci.*, **44**, 112.
90. LATEGAN, P. M. and VAUGHAN, R. H. (1964) *J. Fd Sci.*. **29**, 339.
91. SHIBASAKI, I. and IIDA, S. (1968) *J. Fd Sci. Technol.*, **14**, 447.
92. SHIBASAKI, I. and TSUCHIDO, T. (1973) *Acta Aliment.*, **2**, 237.
93. SHIBASAKI, I. (1969) *J. Ferment. Technol.*, **47**, 167.
94. FAY, J. P. and FARIAS, R. N. (1976) *Appl. Environ. Microbiol.*, **31**, 153.
95. TRELASE, R. D. and TOMPKIN, R. B. (1976) US Patent, 3 955 005.
96. TURCOTTE, P. and SAHEB, S. A. (1978) *Can. J. Microbiol.*, **24**, 1306.
97. SAHEB, S. A., TURCOTTE, P. and PICARO, B. (1978) *Can. J. Microbiol.*, **24**, 1321.
98. LEVIN, R. E. (1967) *J. Milk Fd Technol.*, **30**, 277.
99. NICKERSON, J. T. R. and DARACK, J. R. (1978) US Patent, 4 076 850.
100. UGURBIL, K., ROTTENBERG, H., GLYNN, P. and SCHULMAN, R. G. (1978) *Proc. Natl. Acad. Sci. USA.*, **75**, 2244.
101. UGURBIL, K., BROWN, T. R., DEN HOLLANDER, J. A., GLYNN, P. and SCHULMAN, R. G. (1978) *Proc. Natl. Acad. Sci. USA.*, **75**, 3742.

Chapter 6

NEW PRESERVATIVES AND FUTURE TRENDS

JAMES L. SMITH

US Department of Agriculture, Philadelphia, USA

and

NICHOLAS D. PINTAURO

Rutgers University, New Brunswick, USA

INTRODUCTION

The formation of nitrosamines in meat products preserved with nitrates and nitrites has initiated studies on levels and efficacy of nitrite as food preservatives and on possible nitrite substitutes. Basic studies on the mechanism and role of nitrites in food preservation and product colour and flavour have led to the allowance of lower safe levels of nitrites in meat products, given new confidence in the use of nitrites, and produced new information on the relationship of nitrites with food ingredients and processing variables. Accordingly, other preservatives which have been abandoned as ineffective or of limited effectiveness because of off-flavour or toxicological reasons should be re-examined or re-evaluated.

Historically, the levels of preservatives added did not reflect precise and critical levels dependent on processing, product and toxicological tolerances. Recent studies indicate synergistic actions occurring between nitrite and common food additives (e.g. cysteine, EDTA, ascorbate) used for sequestering and/or control of oxidation/reduction reactions. In earlier research, interactions of preservative with pH, salts (chlorides and phosphates), or water activity have been studied. These have now been extended to other dependent variables: processing temperature, methods of

137

addition, sequestering agents, other preservatives and reducing agents. New methods in experimental design have enabled investigators to examine these variables.

A special problem with pH-dependent food preservatives exists because, as with water content, it is difficult to determine where pH boundaries exist. Measurements of pH, water content, or salt concentration should be considered on a non-homogeneous basis. Bacterial contamination in whole pieces or portions of food exists in layers, and similar profiles exist also with pH, water content, salt content, and even preservative concentration. Bryan[1] refers to this type of physical boundary as 'interfaces' when the boundaries are different components of a food, such as a cake and its filling, meat emulsion and the protein skin, and crust and crumb of bread, which allows differences in concentration of the essential elements for food preservation (pH, water, salt, preservative). In a two-component food such as a filled cupcake, the cake portion can be above pH 4·5 and the filling below pH 4·5 with differences also in the water activity (A_w). When does such a food become a hazard on the basis of pH and A_w? Bakery products with cream fillings, regardless of the pH of the filling, should be classified as hazardous foods requiring continual refrigeration. Because of the interface and the effects of diffusion or migration in terms of contamination and preservation action, many food products which heretofore have been considered non-hazardous should be re-examined for bacteriological safety.

In an excellent review on the mechanism of anti-microbial action of food preservatives, Oka[2] concluded that, although much knowledge has been obtained on the inhibition of metabolic processes in microbial cells by chemicals used as food preservatives, the factors that determine the anti-microbial effect remain undetermined. For effective activity, the chemical component must be transferred from the medium into the cell, and preservatives have been classified according to the manner of this transfer process. The effectiveness of preservatives in the absorption group is determined by the amount of preservative absorbed. Another group acts by oxidation of coenzymes within the cell. Both mechanisms are affected by cell permeability. Therefore, the lipophilicity of the compound and the surface activity of the cell membrane are important factors in the determination of food preservative efficiency. Thus, the destruction of cells is dependent on the interaction of the compound with specific sites on the membrane, rupturing the membrane and exposure of the cell components to the environment.[2]

Some research areas remain. What is the mechanism for inhibitory action? Do food preservatives lower heat resistance? Are preservatives absorbed and pass through the cell membrane, or do they react with the membrane?

The use of food preservatives in the 1980s should have much the same status as anti-biotics had in the 1960s when the emphasis and main objectives in research were not screening and identification of new anti-biotics, but rather the development of new, more efficient systems and conditions for the use of existing anti-biotics.

This chapter presents a review on the status of present preservatives in new systems and combinations; discussion on past research in such areas as anti-biotics for food preservation, radiation and special packaging; and the re-evaluation of these methods in the light of recent findings on mechanisms and effectiveness of traditional food preservatives. The contribution of fatty acids, anti-oxidants and plant extracts in food preservation, and new approaches being investigated to integrate packaging with the anti-microbial requirements for perishable foods, are also discussed.

NITRITE—NEW DEVELOPMENTS

Nitrite has had a long history as meat preservative; however, in recent years its use has been under attack because of the production of nitrosamines from the combination of nitrite with secondary amines. Knowledge of the mode of anti-bacterial action of nitrite will enable researchers to design substitutes that will mimic nitrite action against bacteria. The demonstration that iron (Fe^{2+} and Fe^{3+}) antagonises the anti-botulinal action of nitrite may indicate how nitrite behaves against micro-organisms.

Combining nitrite with reducing agents enhanced the activity of nitrite against *Clostridium botulinum* in meat slurries and in perishable canned cured meats.[3-6] Ascorbate, isoascorbate and cysteine were effective in enhancing the anti-bacterial action of nitrite. Thioglycollate had some activity, whereas sodium formaldehyde sulphoxylate, sodium formaldehyde bisulphite, or sodium sulphide possessed no nitrite-enhancing activity. Thus, reducing activity *per se* did not lead to enhancement of anti-botulinal effects of nitrite. None of the compounds that enhanced nitrite activity had any effect against *C. botulinum* in the absence of nitrite.

When isoascorbate and nitrite were present in perishable canned cured pork there was considerable delay in the outgrowth of *C. botulinum* compared to that in cans with nitrite alone. The combination of isoascorbate and nitrite added to perishable canned cured pork hearts gave an unexpected result—there was no inhibition of *C. botulinum* outgrowth.[5] However, when the isoascorbate and nitrite combination was supplemented with EDTA, *C. botulinum* outgrowth was prevented in the canned pork hearts. Pork hearts contain approximately four times the iron that pork ham contains,[7] and it is suggested that the amount of isoascorbate used in the formulation was not enough to bind all of the iron present in canned cured pork hearts unless EDTA was added also.

Since EDTA is a known sequestrant of divalent metallic ions, Tompkin *et al.*[5] postulated that ascorbate, isoascorbate and cysteine, like EDTA, enhance nitrite anti-bacterial action by chelating an essential metal required by *C. botulinum* for outgrowth. However, Morris *et al.*[8] observed that ascorbate (and presumably isoascorbate) was an excellent chelator of Cu^{2+} but bound Fe^{3+} poorly. Therefore, ascorbate enhancement of nitrite anti-botulinal activity may be due to factors other than iron binding.

In later work, Tompkin *et al.*[7,9] showed that addition of iron (Fe^{2+} or Fe^{3+}) to canned perishable cured beef or pork containing nitrite allowed the outgrowth of *C. botulinum*. The addition of Mg^{2+}, Mn^{2+}, or Zn^{2+} did not have a similar effect on nitrite. Thus, a supply of available iron completely abolishes the anti-botulinal activity of nitrite. Unfortunately, the authors did not attempt to reverse the effect of iron by the addition of EDTA.

Tompkin *et al.*[7] postulated that nitric oxide (from nitrite) combined with ferrodoxin (or a similar iron-containing compound) in germinated *C. botulinum* spores and prevented energy metabolism needed for outgrowth. An excess of available iron in the meat product combines with nitrite and prevents its anti-bacterial action. A search for suitable sequestrants for use in meat products could very well result in greatly decreased nitrite levels.

STARTER CULTURE INHIBITORS

It has been known for many years that there are natural inhibitory systems in raw milk. Under certain circumstances when raw milk is added to a food, growth of many bacteria are inhibited. Antibiotic-like and other inhibitory substances produced by various members of the family *Lactobacteriaceae* and extracted from culture media have been identified as diplococcin, nisin and several other anti-biotics of lesser activity.[10] Diplococcin is known to inhibit most streptococci and other gram-positive cocci. Nisin has a broad spectrum of activity including several streptococci groups. Other inhibitory substances found in dairy substrates inoculated with starter cultures include acids (primarily lactic and acetic acids), CO_2, hydrogen peroxide, polypeptides and certain volatile fatty acids.[11] The lactoperoxidase/thiocyanate/H_2O_2 system found in milk is also bactericidal. The hydrogen peroxide is produced by lactic acid bacteria, and the lactoperoxidase and thiocyanate are found in milk.[12]

It is an acceptable practice in the cheese industry to add lactic acid culture to milk when received under conditions that produce only a slight increase in titratable acidity and/or decrease in pH. The object is to inhibit bacterial growth during the holding time in the plant and to condition the milk protein at the same time for coagulation by the clotting enzyme treatment which follows. In cottage cheese production, cream innoculated with *Streptococcus cremoris*, *S. lactis* and *S. diacetilactis*,[13] can be added to cream the curd. This processing step prevents slime formation and increases shelf-life of the final product. Butter cultures have been used for many years to produce desired flavour and to inhibit the growth of various bacteria that cause flavour defects in butter. The culture organisms compete with spoilage organisms for nutrients and oxygen.

There is limited use of starter cultures in other fermented foods, or fermentations are often carried out sometimes by the normal lactic flora found in the foods. Manufacturers of fermented foods other than fermented milk products seem reluctant to change their traditional processes.

In the production of dry and semi-dry fermented sausages, fewer than half of the processors use lactic acid starter cultures. Few of the fermented sausages sold in local supermarkets list starter culture on the label.[14] The reluctance of meat processors to use starter cultures should be overcome because the advantages of using starter cultures are manifold: (1) large numbers of bacteria of|proven lactic acid producing capacity are added; (2) more rapid acid production decreases the chance of undesirable organisms taking over the fermentation; and (3) processing time is decreased, since it is not necessary to allow for the growth of the natural lactic flora to numbers sufficient to initiate fermentation.

In the absence of starter culture, growth of *Staphylococcus aureus* in a thuringer-type sausage was not affected; when fermentation was initiated by a lactic acid starter culture containing *Pediococcus cerevisiae* and/or *Lactobacillus plantarum*, growth of *S. aureus* was prevented.[15] Also, in the absence of starter culture, measurable amounts of staphylococcal enterotoxin A were detected in European dry sausages.[16]

In summer-style sausage, Christiansen *et al.*[17] demonstrated that the mixture of *L. plantarum* and *P. cerevisiae* prevented toxin formation by

C. botulinum in 23 of 25 sausages; addition of 50 ppm nitrite with the starter culture prevented toxin formation completely. Omission of glucose led to large numbers of toxic sausages. It is necessary that the processor add sufficient fermentable carbohydrate to obtain inhibitory concentrations of lactic acid.

Smith *et al.*[18,19] demonstrated that *Salmonella dublin* and *S. typhimurium* were killed more consistently during lebanon bologna and pepperoni processing when starter cultures containing a mixture of *P. cerevisiae* and *L. plantarum* were used to initiate the fermentation. In natural flora-fermented sausages, a heat treatment was necessary to ensure destruction of *Salmonellae*. Thus, sausage fermentations initiated by addition of lactic acid starter cultures were more effective than those initiated with natural flora in preventing growth and toxin formation by food-poisoning bacteria.

Products such as fermented vegetables (cucumbers and sauerkraut) purchased in local supermarkets do not have starter cultures listed on the labels. Although the advertising literature indicates that starter cultures are available for fermented vegetables, they are probably not in common use. Lactic acid starter cultures would be advantageous to the vegetable fermentation industry by decreasing spoilage and the growth of food-poisoning micro-organisms.

Lactic acid bacteria can be added to a variety of meat products and pasteurised liquid whole egg to prevent both the growth of spoilage organisms and food-borne pathogens. Raccach and Baker[20] found that lactic acid bacteria protect pasteurised liquid whole egg from spoilage (the spoilage organisms could be present as a result of post-processing contamination or underpasteurisation), but the starter was not effective in preventing the growth of *S. typhimurium*. However, in cooked, mechanically deboned poultry meat, the same starter culture inhibited the growth of *S. typhimurium*.[21] Thus, the potential effects of lactic acid bacteria cannot be predicted from one food to another.

Bryan[22] reported that staphylococcal food poisoning is found frequently in hams and ham products. Starter cultures may be useful in preventing growth of staphylococci in ham products. *Streptococcus diacetilactis* present in temperature-abused (25°C) ham sandwich spread killed *Staphylococcus aureus*.[23] Bartholomew and Blumer[24] showed that *P. cerevisiae* repressed the growth of normal ham flora. While they did not study the effect of starter culture on *S. aureus*, the authors suggested that addition of lactic acid bacteria might be useful in preventing growth and toxin formation by food-borne pathogens in hams.

The shelf-life of ground beef could be extended by the addition of a variety of lactic acid bacteria.[23,25] Although not directly related to preservation by lactic acid bacteria, the addition of 2 % glucose increased the shelf-life of ground beef by several days.[26] There was no change in the normal flora; the spoilage organisms preferentially used the added sugar, produced a lower pH, and repressed their own growth. When all of the available glucose was utilised, the micro-organisms began to metabolise the nitrogenous compounds of meat, leading to an increased pH, off-odours and surface slime. Addition of at least 2 % glucose increased the low-temperature shelf-life of ground beef from 5 days to 8–10 days.

Studies by Riemann and his co-workers[27] indicate that semi-preserved meats containing radiation-killed *P. cerevisiae* would not be a food-poisoning hazard. Under conditions of temperature abuse, the killed pediococci utilise fermentable carbohydrates (normally added to semi-preserved meats) to produce acid which would prevent the growth of food-borne pathogens. Under refrigeration, the radiation-killed pediococci would be inactive. This technique could be applied to other foods.

In fermented products, lactic acid is the chief agent in the prevention of spoilage and growth of food pathogens; the reason for the protective action of lactic acid bacteria in unfermented products is unclear. Inhibitory substances such as anti-biotics, H_2O_2 and unidentified compounds have been found in lactic acid bacterial culture fluids.[10,11,28] Recent work on the purification of fermentation liquor obtained from the growth of lactic acid bacteria in culture broth indicates that active anti-bacterial substances are present.[29,30] Further work is needed to identify the compounds and to establish whether it would be possible to add them to foods as preservatives.

FATTY ACIDS AND DETERGENTS

The antiseptic and disinfectant properties of soap are considered to be from residual alkalinity. However, the anti-microbial action is still apparent even at low concentrations of soap in the range 5–50 ppm. At these concentrations, the ionisation of the fatty acid is not important. This has interesting implications in food preservation because fatty acids can be used with wide tolerances in foods.

The anti-microbial effects of small amounts of long-chain fatty acids have been documented by Nieman;[31] gram-positive bacteria are generally more susceptible than gram-negative organisms. A common feature of detergents (which include long-chain fatty acids), phenols, quaternary ammonium compounds and polypeptide anti-biotics is their ability to bind to the cell membrane and cause its disruption as the semi-permeable barrier between cell and environment.[32]

Long-chain fatty acids are not bactericidal to the tubercle bacillus at neutral pH but become so under acid conditions. A virulent strain of *Mycobacterium tuberculosis* normally resistant to 0.1 N HCl (pH 1.0) was rapidly killed at the low pH when traces of fatty acids (C_{12}, C_{14}, C_{16}, $C_{18:1}$) were present.[33] Kondo and Kanai[34] tested fatty acids against various species of mycobacteria; long-chain fatty acids (C_{14}, $C_{18:1}$, $C_{19:2}$) were active against all test strains.

With *M. bovis*, the unsaturated long-chain fatty acids inhibited membrane-bound enzyme activity (acid phosphatase and tetrazolium reductase) as well as growth.[34] Oleic acid had only slight mycobactericidal activity at pH 7 but had very strong activity at pH 5.6. Addition of the basic protein protamine to oleic acid at acid pH led to loss of mycobactericidal activity.[34] Further work by Kondo and Kanai[35] indicated that bactericidal effects of long-chain fatty acids was of non-specific nature and was due to the detergent-like action on the cytoplasmic membrane. The difference in sensitivity between gram-positive and gram-negative bacteria was shown to be due to the ability of the former to adsorb long-chain fatty acids.

The growth of gram-positive organisms belonging to the genera *Bacillus*, *Streptococcus*, *Staphylococcus*. *Micrococcus* and *Clostridium* was inhibited by sodium laurate or sodium linolenate (0.5 mM or less at pH 7.4 in bacterial media). Bacilli were inhibited by lower concentrations of the long-chain fatty acid than were the other genera. Gram-negative cells were not affected.[36] Lauric acid was the most effective saturated fatty acid tested but was not as effective as C_{18} unsaturated fatty acids. Calcium and magnesium ions, cholesterol and ergocalciferol reversed the fatty acid inhibition of growth. Kabara *et al.*[37,38] studying gram-positive cocci (streptococci, micrococci and pneumococci), found that lauric acid was the most active saturated long-chain fatty acid. The monoglyceride, 1-monolaurin, was more active in preventing bacterial growth than was the free acid; 1,3-dilaurin or the triglyceride had no effect.

Utilising *B. subtilis*, Sheu and Freese[39] showed that the concentration of fatty acid required to inhibit growth increased with decreasing molecular weight (pH 7 in bacterial media); 0.5 mM $C_{18:2}$ gave as much inhibition as did 100 mM C_4. Freese *et al.*[40] reviewed the anti-microbial properties of lipophilic food additives. Most lipophilic food preservatives prevent bacterial growth by inhibiting the transport of substrate molecules into cells. Saturated long-chain fatty acids uncouple transport of substrate and oxidative phosphorylation from the electron transport system and inhibit cellular uptake of amino acids, organic acids and phosphate through the membrane. Unsaturated long-chain fatty acids also appear to inhibit the electron transport system of the cell.

At pH 6·0, lauric acid was effective against *B. megaterium* and *M. lysodeikticus*; with increasing pH, a higher concentration of the fatty acid was necessary to show growth inhibition (0·03 mM at pH 6 as compared to 0·3 mM at pH 8). The anti-bacterial effect of linoleic acid decreased with increasing pH but to a lesser degree than that of lauric acid.[41] The uptake of both lauric and linoleic acids by *B. megaterium* and *M. lysodeikticus* was governed by pH also. The amount of fatty acid uptake decreased as the pH increased and reflected the decrease in bactericidal activity. Although *Pseudomonas phaseolicola* was resistant to the inhibitory action of long-chain fatty acids, it did adsorb these fatty acids. Protoplasts of *B. megaterium* had a greater uptake of the long-chain fatty acids than had whole cells; thus, removal of the cell wall allows greater exposure of the cell membrane to the fatty acids. Galbraith and Miller[42] showed that addition of bovine serum albumin antagonised the lytic action of lauric and linoleic acids on protoplasts of *B. megaterium*. The added protein may have absorbed the fatty acids, removing them from the cellular environment, and prevented cell lysis.

Galbraith and Miller[42] also studied the degree of protoplast lysis by measurement of leakage or release of purines, pyrimidines, proteins and peptides. The molar concentration of capric acid (C_5) necessary to produce lysis was 10 times greater than that of lauric acid or linoleic acid.

Long-chain fatty acids stimulated oxygen uptake by *B. megaterium* and *M. lysodeikticus* at bactericidal concentrations but inhibited oxygen uptake at high concentrations.[43] Protoplasts were more susceptible than whole cells to inhibition. Oxygen uptake by whole cells of *P. phaseolicola* was not inhibited by lauric or linoleic acids but spheroplasts were susceptible to the action of the fatty acids.[43] Long-chain fatty acids inhibited the uptake of glutamic acid and lysine by *B. megaterium* and *M. lysodeikticus*; the inhibiting effect was reversed by Ca^{2+}. The uptake of lysine was prevented by lauric and linoleic acids in *C. welchii*. High levels of fatty acids were necessary to inhibit lysine or glutamate uptake in *P. phaseolicola* and *Escherichia coli* but spheroplasts of the gram-negative cells were just as sensitive as were gram-positive whole cells. Long-chain fatty acids (C_{12}, C_{14}, C_{18}, $C_{18:2}$) caused leakage of glutamate from cells of *B. megaterium* preloaded with radioactive glutamate. Thus, the work of Galbraith and Miller indicates that long-chain fatty acids act on the cytoplasmic membrane (site of O_2 uptake and transport of amino acids) probably involving the uncoupling of energy systems.

Sheu and Freese[44] examined the lack of inhibition of most gram-negative bacteria by long-chain fatty acids. Growth, amino acid transport and O_2 consumption in *E. coli* and *Salmonella typhimurium* were inhibited by short-chain fatty acids (C_{2-6}) but not by medium- or long-chain fatty acids (C_{10-18}). The resistance is not correlated with the ability of the cells to metabolise fatty acids but is due to the nature of the lipopolysaccharide (LPS) layer which surrounds gram-negative bacteria. The LPS layer prevents accumulation of long-chain fatty acids on the inner cell membrane (site of amino acid transport). Inhibition of the growth of an *E. coli* mutant defective in the LPS layer occurred with levels of decanoate that were not effective with normal *E. coli*. However, the LPS defective mutant was not inhibited by oleate. Addition of EDTA to the gram-negative organisms removed part of the LPS layer, and the cells then became sensitive to long-chain fatty acids (C_{10} and $C_{18:2}$). Fay and Farias[45] also showed that spheroplasts of *E. coli* were sensitive to methyldecanoate at levels that had no effect on whole cells. Sheu and Freese[44] suggested that the addition of EDTA along with lipophilic food preservatives would increase the effectiveness of those inhibitors against gram-negative bacteria.

Kato and Arima[46] found that lauric acid inhibited growth of *E. coli* more effectively than did other fatty acids ($C_{10}-C_{18:3}$) tested. The authors synthesised the water-soluble sucrose mono-ester of lauric acid, which was inhibitory at $100 \mu g/ml$ (1 mg/ml lauric acid showed partial inhibition only). However, the sucrose ester was only bacteriostatic, and growth of *E. coli* occurred after a long lag. Kabara[47] found that sucrose laurate was no more active than lauric acid against members of *Streptococcus* group D or *Staphylococcus epidermis*.

Relatively low concentrations of long-chain fatty acids are inhibitory to gram-positive bacteria in bacterial media. Unfortunately, the long-chain fatty acids are not effective against gram-negative bacteria. The inhibitory effect of the long-chain fatty acids is more marked at lower pH, and the inhibitory action is reversed by additions of proteins or by ions such as Ca^{2+}.[42,43] Gram-negative cells are more affected by short-chain fatty acids.[44] It might appear feasible to formulate foods with short-chain fatty acids to lower the pH; addition of long-chain fatty acids along with the short-chain fatty acids would inhibit both gram-positive and gram-negative bacteria. However, the effect of such additives on the flavour and physical characteristics of food must be considered. The precise composition and the physical and chemical form or condition of the fatty acid additive can be well defined and controlled by sophisticated analytical methods. Simple dispersion of a fatty acid in a food cannot be effective for food preservation purposes. It may be critical, for example, that the fatty acid exist as unimolecular films so that they form boundaries or cover large surface areas. Also, the fatty acid may exist in some complex form with proteins, carbohydrates, or some other forms of lipid material.

ANTI-OXIDANTS AS INHIBITORS

Butylated hydroxyanisole (BHA) and butylated hydroxytoluene (BHT) are frequently added to foods because of their anti-oxidant properties; however, these hindered phenols are similar to other phenols in possessing anti-microbial activity.

Chang and Branen[48] reported that the growth and aflatoxin production of *Aspergillus parasiticus* were prevented by 250 ppm BHA in growth medium. Fung et al.[49] found that growth of toxin formation were inhibited in six strains of *A. flavus* by BHA (0·005–0·02 g/petri dish). BHT at similar concentrations had no effect.

Staphyloccus aureus growing in nutrient broth was quite sensitive to BHA: 150–200 ppm was bactericidal. The bactericidal level of BHA for enteropathogenic *E. coli* was 400 ppm, but such levels delayed the growth of *Salmonella typhimurium* only slightly.[48]

In a nutrient medium, 50 ppm BHA rapidly killed *Vibrio parahaemolyticus*. However, in a sterile homogenate of blue crabmeat, 400 ppm was necessary to exert a bactericidal effect. Robach et al.[50] suggested that the decrease in the anti-bacterial effect of BHA in crabmeat was due to partial inactivation of the BHA by oxidised lipids. The anti-bacterial effect of BHA in foods, therefore, may depend on the content and type of lipids present.

Klindworth et al.[51] tested BHA against *Clostridium perfringens* in fluid thioglycollate medium and found that 200 ppm inhibited the growth of the organism. BHA was stable to autoclaving and reacted synergistically with nitrite, sorbic acid and parabens. Unfortunately, the addition of lipid to the medium drastically reduced the effectiveness of BHA against *C. perfringens*.

Although BHA shows anti-microbial activity in culture media, its utility as a food preservative has not been demonstrated. The decrease in activity found when lipids are added suggests that the anti-bacterial activity and the anti-oxidant effect are linked and that BHA which reacts with lipids is no longer available for anti-bacterial activity. Alternately, lipid inactivation of the anti-bacterial properties of BHA may be due to solubilisation of the anti-oxidant in the lipid, thus making it unavailable for action upon micro-organisms.

SEQUESTRANTS

Sequestrants (chelating agents) react with metals to form complexes that in many cases effectively bind the metal ions and remove them as reactants or catalysts in reactions that cause deterioration of the food. Many sequestrants occur naturally in foods; these include polycarboxylic acids (oxalic, succinic), hydroxycarboxylic acids (citric, malic, tartaric), polyphosphoric acid (ATP), amino acids (glycine, leucine, cysteine) and various macromolecules (porphyrins, peptides and proteins). Metal chelating agents can act alone or synergistically with other compounds.[52]

There are biological sequestrants that have been shown to have anti-bacterial activity. Micro-organisms, like other biological forms, require metallic ions (especially iron) for growth and metabolism. For example, conalbumin, an iron-binding protein found in egg-white, has inhibited the growth of various micro-organisms and such inhibition is reversed by iron.[53] Washing eggs in water containing iron will accelerate spoilage.[54] Thus, the iron-sequestering action of conalbumin protects eggs from microbial deterioration. Lactoferrin, the milk iron-binding protein, inhibits the growth of E. coli in vitro, and the effect is abolished by the addition of iron.[55,12] There are no reports concerning the possibility that lactoferrin protects milk or milk products from microbial spoilage.

Higher organisms can obtain iron from the foods they ingest, but bacteria must extract iron directly from the environment. Since iron is insoluble, the concentration available for bacterial growth is extremely low. Therefore, bacteria produce small organic molecules (siderochromes or siderophores) that facilitate transport of iron into the cell.[56] Bacteria also produce other chelating agents to facilitate uptake of metallic ions other than iron. Weinberg[57] suggested that the addition of microbial sequestrants (or analogues thereof) to food products may be useful in the prevention of microbial spoilage or to prevent the growth of food-borne pathogens. Feeding mice a basal diet supplemented with enterobactin (an iron-binding agent produced by a number of gram-negative species) protected them against Salmonella typhimurium infection.[58]

Fresh meats spoil quickly at low temperatures because of the growth of

Pseudomonas species. Since these bacteria require iron for growth and obtain it from the meat substrate, the addition of the specific microbial sequestrant to bind iron could prevent the bacterial growth that leads to meat spoilage.

The possibilities for control or inhibition of bacterial growth in specific foods by control and manipulation of iron availability by use of synthetic or natural sequestrants remain to be studied.

CONTROLLED ATMOSPHERES

The aerobic meat spoilage *Pseudomonas* species are inhibited by use of carbon dioxide in storage and packaging. Carbon dioxide atmospheres for storage of fruits and vegetables retard respiration, ripening and the growth of bacteria. The systems used for fruits and vegetables require carbon dioxide generators and air-tight bins or containers in refrigerated storage areas. Significant progress has been made with in-package controlled atmosphere systems in which films are used with selective permeabilities of oxygen and carbon dioxide.[59] In such a system there is a gradual accumulation of carbon dioxide in the package due to respiration of the fruit or vegetable product. The ratio of carbon dioxide to oxygen is maintained to inhibit growth of anaerobic bacteria which produce off-flavours.

Benedict *et al.*,[60] utilising a mixture of sodium bicarbonate and citric acid, developed an in-package system for the generation of CO_2 in packaged meat. The gas is produced as the moisture within the package increases and effectively retards the growth of principal meat spoilage organisms. Other studies have been conducted to evaluate the effect of various gas atmospheres on microbial growth, pH and colour of fresh meat during storage.[61] Samples were stored in barrier bags which were evacuated of air and flushed with gas and sealed. Samples stored in CO_2 had significantly lower aerobic bacterial counts than N_2, O_2, or air, while anaerobic bacterial counts did not increase during 27 days of storage.

The risk in vacuum packaging in former years was in the maintenance of low oxygen atmospheres because of weaknesses in the package seal or in the package material itself. These problems have been generally eliminated by new progress in packaging technology; for example, the hypobaric transport and storage of fresh foods utilises a precisely controlled combination of low pressure, low temperature and high humidity to achieve extended shelf-life.[62] Anaerobic bacterial growth, which would not be inhibited in vacuum packages, is also a problem. This entire area of technology of gaseous atmospheres for preservation by anti-bacterial action should be evaluated for different product classes (meat, fish, dairy and other protein-containing foods). This objective has been incidental in earlier studies because of the more urgent demands for retardation of staling, rancidity, flavour loss, and other storage effects of chemical deterioration.

INHIBITION BY REDUCED WATER ACTIVITY

Sugars and sugar alcohols are being used in special food applications to reduce water activity and thus inhibit microbial growth. It is a common practice to add salts, sugars and glycols to foods for this purpose, and such ingredients may be self-limiting because of saltiness, sweetness and bitterness. Novel approaches to manipulate the chemical and physical structure of foods and the distribution of water in the food have been studied. The outgrowth of these studies probably will be a new generation of technology and product applications that will allow storage of foods without refrigeration or thermal processing; minimal changes in flavour, colour and texture will occur in these foods compared with their conventional or traditional counterparts.

The lowered water activity is probably necessary to make water unavailable for solubilisation and transport of nutrients required for the growth of micro-organisms. One approach to accomplish this objective is to reduce the amount of immobilised water held in the tissue structure of the food by the principle of hydrostatic equilibrium or osmotic drying. The piece of vegetable or meat is placed in a high solids medium such that a hydrostatic pressure differential is developed; the water inside the food piece will tend to migrate toward the concentrated solution. Depletion of water inside the cell then destroys the nutritional balance of water-soluble nutrients for the growth of micro-organisms. At the same time, depletion of water causes concentration of salts, which will denature the protein components of micro-organisms. Other approaches are as follows.

Enzymes
Enzymes effective in the hydrolysis of plant gums and breakdown of macromolecules *in situ* to effect localised hydrostatic differences in the intracellular spaces can be added to foods. This approach will provide a shorter pathway to transfer water from areas of low solids (higher water

content) to areas of higher solids content. Lower-molecular-weight species or breakdown products from complex polysaccharides and proteins are converted to simpler soluble compounds. In the lower-molecule-weight form, hydration as well as solubility will be enhanced, giving greater concentration differentials and finally more effective hydrostatic action.

Hydrogen Bonding

Hydrogen-bond breakers, such as urea or quanidinum chloride which would accelerate the dissociation of water from polar groups of protein and carbohydrates, should be studied. The hydrogen-bond breaking mechanism may help to produce unequal moisture regions, allowing hydrostatic forces to work.

Water of Hydration

Water that is chemically free is in the optimum state for chemical and bacteriological activity. Increases in the water of hydration may have larger effects on the distribution and balance of free water and create unfavourable conditions for micro-organisms.

LIQUID SMOKE

One of the difficulties in assessing the anti-bacterial properties of smoke is that heating and drying effects are also part of the smoking process. However, previous research over the years indicates that smoking imparts anti-bacterial and anti-oxidant properties to foods.

Liquid smoke preparations can be diluted and added to bacteriological media in such a way that a quantitative measure of the anti-bacterial properties of the smoke can be made. The factors of heating and drying are eliminated. Unfortunately, liquid smoke preparations are selected for their flavour qualities rather than for their anti-bacterial properties and thus have not been studied extensively from a bacteriological viewpoint. The composition of the 'smoke' flavour fraction of CharSol C-10 has been described by Fiddler et al.[63,64] The ether-soluble fraction that has the smoke flavour contained primarily phenols and carbonyls. The anti-bacterial potential of the fraction was not studied.

Handford and Gibbs[65] prepared smoked water concentrates by smoking water in plastic casings. The resultant smoke liquids were tested against various micro-organisms. They found that 22 of 39 members of the family *Lactobacteriaceae* were not inhibited by the smoke preparations, whereas only 2 of 26 catalase-positive cocci were resistant. *Staphylococcus aureus* strains were strongly inhibited. Handford and Gibbs also showed that smoking tended to accelerate the shift from a predominant catalase-positive flora (micrococci) to a lactic acid flora during storage of vacuum-packed sliced bacon. Thus, the data obtained with the bacon studies were substantiated by the model system liquid smoke studies with bacterial media.

Houben[66] tested five different smoke preparations against micro-organisms growing in bacteriological media. One preparation was strongly bacteriostatic, two had moderate activity, and two were inactive. *Bacillus* species and *Microbacterium* species were quite sensitive to liquid smoke, but *E. coli* and members of the family *Lactobacteriaceae* were resistant. Two of the preparations were tested in meat suspensions by Houben[67] against *B. cereus*, *B. subtilis*, *E. coli*, *S. aureus*, *Streptococcus faecium* and one species each of *Lactobacillus*, *Microbacterium* and *Micrococcus*. CharSol EZC85 inhibited growth of *S. aureus* (at about the 1·3 % level), but the other micro-organisms were not affected by 2 % (the highest level used). Naarden IM142 inhibited the bacilli and *Microbacterium* at the 0·15 % level or less and *Lactobacillus* at 0·7 %, but did not affect the other bacteria. Houben's work indicated that anti-bacterial potential among various smoke preparations may differ greatly.

If liquid smoke preparations are selected for both flavour and anti-bacterial activity, then liquid smoke can play a role in preserving those foods that are normally smoked.

NATURAL PHENOLS AND RELATED COMPOUNDS

Over the years various researchers have conducted screening-type investigations on the anti-bacterial activity of plant extracts. In one such investigation, Powers[68] studied anthocyanin, leucoanthocyanins and phenol acids. He tested 24 compounds for their effect on respiration or reproduction of bacterial cells. When glucose was present, anthocyanin and leucoanthocyanin inhibited *E. coli*, *Salmonella typhosa*, *Aerobacter aerogenes*, *Proteus vulgaris* and several other bacterial species. In the absence of glucose, the bactcrial cells metabolised these plant pigments. *Para*-hydroxybenzoic acid, gallic acid and vanillic acid inhibited *E. coli*, *A. aerogenes* and *P. vulgaris* in human urine.

Extracts prepared from certain wood species and containing phenolic neoflavanoids (and isomers of cinnamylphenols) exhibited anti-bacterial

Developments in Food Preservation — 1
Edited By R. H. Tilbury.
1980.

NEW PRESERVATIVES AND FUTURE TRENDS 153

activity against a variety of gram-positive bacteria, yeasts and moulds. Jurd et al.[69] reported that the minimal inhibitory concentration compared favourably with several synthetic anti-microbial agents. Extracts were prepared by Smale et al.[70] from fresh leaves, stems and flowers of 125 plant species and varieties for measurement of anti-microbial activity, and 21 of 45 active extracts showed sufficient activity and potency to warrant further study. Bloomfield[71] studied the mechanism of action of phenolic anti-bacterial agents and concluded that these compounds disrupted active transport in the cell membrane. The phenolic compounds accelerate proton translocation across the membrane. Bloomfield obtained experimental evidence for this conclusion by studying the bacteriostatic action of Fentichlor against S. aureus and E. coli.

IRRADIATION

Extensive research and development work has been completed on the use of ionising radiation as a method for food preservation. The subject has been thoroughly reviewed by Urbain.[72] In addition, the *Proceedings of the International Symposium on Food Preservation by Irradiation* are available from the International Atomic Energy Agency.[73] The subject areas covered at this symposium were: Control of Microbial Spoilage; Chemical Changes; Toxicological Studies; Public Health and Consumer Acceptance; Economics and Energy Aspects; Irradiation Facilities.

It is generally known that low doses of ionising radiation, in the order of 10^4 roentgen equivalent physical (REP), results in a significant increase in the shelf-life of many perishable food products. At such low doses, many bacterial species that cause spoilage survive. For more effective pasteurisation, doses of 10^6 REP must be used. With fresh meat, however, this treatment allows only a moderate extension of shelf-life. Higher doses cause changes in the organoleptic properties of the food, particularly flavour and colour. Radiation treatment combined with other preservatives has been highly effective. In one such example, Wierbicki et al.[74] reported that it is possible to reduce the levels of nitrite and nitrates in cured meats and still have a microbiologically safe product. In irradiation-sterilised cured meats, nitrite can be reduced to levels needed only for colour and flavour. However, small amounts of nitrate must be added to prevent fading of cured meat colour. In irradiation-sterilised ham and corned beef, the total added nitrate and nitrite should be 75 ppm (1/3 to 1/2 nitrite). No nitrosamines were found in fully cooked smoked hams after radiation treatment. In radiation-sterilised prefried bacon, a 1:1 mixture of nitrate and nitrite at 50 ppm gave a satisfactory product. After additional frying to crispness, no nitrosopyrrolidine was found.

Radiation sterilisation is probably the most underutilised method for food preservation. However, three problem areas are associated with irradiation for food preservation: (1) chemical changes which adversely affect colour, flavour and consumer acceptability of the food; (2) the toxicological status of the compounds formed by irradiation treatment; and (3) high cost for irradiation processing and marketing difficulties. Up to the present, there are no product advantages of irradiated foods that could justify higher consumer costs. A joint committee of the Food and Agricultural Organisation/International Atomic Energy Agency/World Health Organisation accepted the concept of food irradiation as a 'food process' rather than a 'food additive'. In addition, they approved several irradiated food items, including chicken, cod and redfish. Some experts are advocating the approval of classes of food because of obvious similarities in their photochemistry.

Preservation methods for frozen foods, thermally processed foods and dehydrated foods are adequate in terms of costs, convenience, consumer acceptance and shelf-life. The most prominent promoter and developer of irradiated foods has been the US Quartermaster Corps. If in the future the marketing and consumer requirements change for different types of long-term shelf-life sterilised foods or for short-term non-refrigerated perishable foods, re-evaluation of present data or new research may successfully overcome the difficulties of consumer acceptance, high costs and possible toxicology of irradiated foods.

ANTI-BIOTICS

The original series of testing of anti-biotics as anti-microbial agents for food preservation in the 1950s included chlortetracycline, oxytetracycline, nisin, pimaricin and nystatin.[75] One of the general conclusions, based on the initial series of studies, is that anti-biotics exhibit selective anti-microbial activity; some are active against many gram-positive bacteria, some are active against gram-negative, and only a few show broad-spectrum activity.[75] One of the more important characteristics of anti-biotics is that they are not influenced by pH. The preservation action is static, and anti-biotic material must be continuously present.

In a typical product application to extend the shelf-life of fresh ground beef, Kohler *et al.*[76] used chlortetracycline in combination with either nystatin of myprozine. The results in Table 1 are based on storage tests of meat patties, tray packed and overwrapped with cellophane.

In another application, fresh fish was dipped for approximately 20 seconds in a water solution containing 10 ppm chlor- or oxytetracycline. A chelating agent such as EDTA was used to maintain the preservative action of the anti-biotic.[77]

TABLE 1

EFFECT OF ANTI-BIOTICS ON THE SHELF-LIFE OF MEAT PATTIES

Group	Treatment before storage	Spoilage (days)	Discoloration (days)
1	Control (untreated)	7	5
2	Chlortetracycline hydrochloride 3 ppm	11	13
3	Chlortetracycline hydrochloride 3 ppm plus 10 ppm nystatin	22 +	20
4	Chlortetracycline hydrochloride 3 ppm plus 10 ppm myprozine	22 +	22

It has been concluded that the use of anti-biotics, clinical or non-clinical, in situations where they can enter the food chain should be discouraged. Micro-organisms easily become resistant to anti-biotics and thus make it necessary to add more anti-biotic or use a different anti-biotic system.

Anti-biotic resistance, especially among the *Enterobactereaceae*, is plasmid mediated. Plasmids are extrachromosomal elements that behave like auxiliary chromosomes and have the ability to promote genetic transfer by conjugation. A plasmid is generally named for the particular property that is specifically its main characteristic. All plasmids are replicating genetic elements which direct their own replication and segregation during cellular division. These plasmids, also called resistance transfer factors (R-factors), have been found in most serotypes of *E. coli*, all four species of *Shigella*, and the Alkescens–Dispar group, many of the *Salmonella*, *Arizona*, *Citrobacter*, all species of *Proteus*, and members of the genera *Providencia*, *Klebsiella*, *Enterobacter*, *Serratia*, *Aeromonas*, *Pseudomonas*, *Yersinia*, as well as others.[78] Plasmids can be transferred between strains of the same species, between species of the same genus, and between genera. In addition to anti-biotic resistance, other phenotypic characteristics are plasmid mediated. Biochemical characteristics, such as sugar fermentation, toxin formation, specific antigens, haemolysins and mucosal adhesion have been associated with plasmids.[78,79] These factors, too, can be passed from one genus to another and thus make accurate identification—so important in food microbiology—virtually impossible.

Often, more than one character is transmitted at a time; multiple drug resistance is common.[80] Wachsmuth *et al.*[81] demonstrated that the plasmid mediation and transfer of heat-stable enterotoxin production and multiple drug resistance in enteropathogenic *E. coli* was responsible for a hospital outbreak. Concurrent transfer of both multiple drug resistance and enterotoxin production into *E. coli* K-12 was demonstrated also. The transfer of both drug resistance and toxin formation could occur between enteropathogenic *E. coli* and other coliforms (or even other genera) found in foods. Thus, because of the intergeneric transfer of plasmids and the possibility of multiple drug resistance, an *E. coli* resistant to anti-biotics present in foods could transfer such characters to a spoilage organism or a pathogen found in foods.

NEW FOOD ADDITIVES

Pharmaceutical and food companies have been active in the identification, development and testing for safety of new compounds that will preserve foods to avoid spoilage, illness and economic loss. Many of these compounds are described in the patent literature and some are listed in Table 2. Information on the food status of these compounds is not known

TABLE 2
PARTIAL LIST OF CHEMICAL PRESERVATIVES IN PATENT LITERATURE[82]

Compound	Food product	Company	US Patent
Thioisomaltol	Fruits	General Foods	3 695 899 (1972)
Imidazoline	Beer	not assigned	3 440 056 (1969)
Lauric diethanol amide	Beer	not assigned	3 440 057 (1969)
Benzohydroxamic acid	General	Chas. Pfizer	3 446 630 (1969)
5-Aminohexahydro-pyrimidine	Fish	Warner-Lambert	2 963 374 (1960)
t-Butyl-hydroperoxide	Fish	Atlantic Richfield	3 622 351 (1971)
Phosphate peroxide	Fish	Takeda	3 545 982 (1970)
Chlorobromo dimethyl hydantoin	Dairy	Genhal	3 499 771 (1970)

by the authors. Enquiries on this subject should be directed to the companies involved.

OUTLOOK

The development and acceptance of new preservatives face a bleak future. Regulatory agencies will take a hard and forbidding look at new formulations. The prohibitive costs to test the safety of new compounds will discourage development by commercial concerns. Consequently, the emphasis in the future will probably be on new uses for old compounds and methods.

For example, lactic acid starter cultures have increased the shelf-life of liquid whole egg, mechanically deboned meats, and ground beef with only minor pH changes; therefore, the traditional purpose of starter cultures (production of large amounts of lactic acid quickly) does not appear to be operating here. Research is needed to clarify the mechanism by which lactic acid bacteria increase the shelf-life of meat products and the possible extension of the use of starters to other food systems.

Long-chain fatty acids and sequestrants are known bacterial inhibitors but are not used in foods to prevent bacterial growth. The bacterial membrane is sensitive to attack by fatty acids; this fact could be utilised to protect food systems from spoilage and pathogenic bacteria. Metal ions—especially iron—are needed for bacterial growth. Addition of purified chelating agents isolated from micro-organisms could play an important role in the extension of the shelf-life of foods. Addition of both a fatty acid and a sequestrant may provide a combination that spoilage and pathogenic bacteria could not overcome. Combinations containing sorbic acid and sodium acid pyrophosphate with or without nitrite were effective in preventing botulinal outgrowth [83] Many combinations are possible with known food preservatives which will prevent bacterial growth effectively.

Controlled atmospheres coupled with a small decrease in water activity would provide conditions which would prevent the growth of many bacteria.

Radiation offers a preservative effect that can have minor effects on food quality and yet offer major protection against bacteria. If the human safety problem can be solved, radiation may well be the preservative of the future.

The food preservatives of the 1980s will probably not be new and unusual compounds but rather will be new and unusual combinations of old and tried compounds and methods.

REFERENCES

1. BRYAN, F. L. (1976) *J. Milk Fd Technol.*, **39**, 289.
2. OKA, S. (1964) 'Mechanism of antimicrobial effect of various food preservatives', *Fourth International Symposium on Food Microbiology*, N. Molin (Ed.), Almqvist and Wiksell, Stockholm.
3. JOHNSTON, M. A. and LOYNES, R. (1971) *Can. Inst. Fd Technol. J.*, **4**, 179.
4. ASHWORTH, J. and SPENCER, R. (1972) *J. Fd Technol.*, **7**, 111.
5. TOMPKIN, R. B., CHRISTIANSEN, L. N. and SHAPARIS, A. B. (1978) *J. Fd Sci.*, **43**, 1368.
6. TOMPKIN, R. B., CHRISTIANSEN, L. N. and SHAPARIS, A. B. (1978) *Appl. Environ. Microbiol.*, **35**, 59.
7. TOMPKIN, R. B., CHRISTIANSEN, L. N. and SHAPARIS, A. B. (1978) *J. Fd Technol.*, **13**, 521.
8. MORRIS, S. G., MYERS, J. S., JR., KIP, M. L. and RIEMENSCHNEIDER, R. W. (1950) *J. Am. Oil Chem. Soc.*, **27**, 105.
9. TOMPKIN, R. B., CHRISTIANSEN, L. N. and SHAPARIS, A. B. (1979) *Appl. Environ. Microbiol.*, **37**, 351.
10. BABEL, F. J. (1977) *J. Dairy Sci.*, **60**, 815.
11. SPECK, M. L. (1972) *J. Dairy Sci.*, **55**, 1019.
12. REITER, B. (1978) *J. Dairy Res.*, **45**, 131.
13. ELLIKER, P. R., SANDINE, W. E.. HAUSER, B. A. and MOSELEY, W. K. (1964) *J. Dairy Sci.*, **47**. 680.
14. Task Force Report (October 1977) *The staphyloccal enterotoxin problem in fermented sausages*, Food Safety and Quality Service, US Department of Agriculture.
15. DALY, C., LACHANCE, M., SANDINE, W. E. and ELLIKER, P. R. (1973) *J. Fd Sci.*, **38**, 426.
16. NISKANEN, A. and NURMI, E. (1976) *Appl. Environ. Microbiol.*, **31**, 11.
17. CHRISTIANSEN, L. N., TOMPKIN, R. B., SHAPARIS, A. B., JOHNSTON, R. W. and DAUTTER, D. A. (1975) *J. Fd Sci.*, **40**, 488.
18. SMITH, J. L., PALUMBO, S. A., KISSINGER, J. C. and HUHTANEN, C. N. (1975) *J. Milk Fd Technol.*, **38**, 150.
19. SMITH, J. L., HUHTANEN, C. N., KISSINGER, J. C. and PALUMBO, S. A. (1975) *Appl. Microbiol.*, **30**, 759.
20. RACCACH, M. and BAKER, R. C. (1979) *J. Fd Sci.*, **44**, 90.
21. RACCACH, M. and BAKER, R. C. (1978) *J. Fd Prot.*, **41**, 703.
22. BRYAN, F. L. (1976) 'Staphyloccus aureus', *Food microbiology: public health and spoilage aspects*, M. P. Defigueiredo and D. F. Splittstoesser (Eds.), Avi Publishing Co., Westport, Conn., USA, pp. 13–128.
23. DALY, C., SANDINE, W. E. and ELLIKER, P. R. (1972) *J. Milk Fd Technol.*, **35**, 349.
24. BARTHOLOMEW, D. T. and BLUMER, T. N. (1977) *J. Fd Sci.*, **42**, 498.
25. REDDY, S. G., HENRICKSON, R. L. and OLSON, H. C. (1970) *J. Fd Sci.*, **35**, 787.
26. SHELEF, L. A. (1977) *J. Fd Sci.*, **42**, 1172.
27. RIEMANN, H., LEE, W. H. and GENIGEORGIS, C. (1972) *J. Milk Fd Technol.*, **35**, 514.
28. GILLILAND, S. E. and SPECK, M. L. (1977) *J. Fd Prot.*, **40**, 820.

29. BRANEN, A. L., GO, H. C. and GENSKE, R. P. (1975) *J. Fd Sci.*, **40**, 446.
30. PULUSANI, S. F., RAO, D. R. and SUNKI, G. R. (1979) *J. Fd Sci.*, **44**, 575.
31. NIEMAN, C. (1954) *Bacteriol. Rev.*, **18**, 147.
32. HAMILTON, W. A. (1968) *J. Gen. Microbiol.*, **50**, 441.
33. HART, P. D'ARCY, LOVELOCK, J. E. and NASH, J. (1962) *J. Hyg. Camb.*, **60**, 509.
34. KONDO, E. and KANAI, K. (1972) *Japan J. Med. Sci. Biol.*, **25**, 1.
35. KONDO, E. and KANAI, K. (1976) *Japan J. Med. Sci. Biol.*, **29**, 25.
36. GALBRAITH, H., MILLER, T. B., PATON, A. M. and THOMPSON, J. K. (1971) *J. Appl. Bacteriol.*, **34**, 803.
37. KABARA, J. J., SWIECZKOWSKI, D. M., CONLEY, A. J. and TRUENT, J. P. (1972) *Antimicrob. Agents Chemother.*, **2**, 23.
38. KABARA, J. J., VRABLE, R. and LIE KEN JIE, M. S. F., (1977) *Lipids*, **12**, 753.
39. SHEU, C. W. and FREESE, E. (1972) *J. Bacteriol.*, **111**, 516.
40. FREESE, E., SHEU, C. W. and GALLIERS, E. (1973) *Nature*, **241**, 321.
41. GALBRAITH, H. and MILLER, T. B. (1973) *J. Appl. Bacteriol.*, **36**, 635.
42. GALBRAITH, H. and MILLER, T. B. (1973) *J. Appl. Bacteriol.*, **36**, 647.
43. GALBRAITH, H. and MILLER, T. B. (1973) *J. Appl. Bacteriol.*, **36**, 659.
44. SHEU, C. W. and FREESE, E. (1973) *J. Bacteriol.*, **115**, 869.
45. FAY, J. P. and FARIAS, R. N. (1977) *J. Bacteriol.*, **132**, 790.
46. KATO, A. and ARIMA, K. (1971) *Biochem. Biophys. Res. Commun.*, **42**, 596.
47. KABARA, J. J. (1975) *J. Cosmet. Perfum.*, **90**, 21.
48. CHANG, H. C. and BRANEN, A. L. (1975) *J. Fd Sci.*, **40**, 349.
49. FUNG, D. Y., TAYLOR, S. and KAHAN, J. (1977) *J. Fd Safety*, **1**, 39.
50. ROBACH, M. C., SMOOT, L. A. and PIERSON, M. D. (1977) *J. Fd Prot.*, **40**, 549.
51. KLINDWORTH, K. J., DAVIDSON, P. M., BREKKE, C. J. and BRANEN, A. L. (1979) *J. Fd Sci.*, **44**, 564.
52. FURIA, T. E. (1973) 'Sequesterants in foods', *CRC Handbook of Food Additives*, 2nd edn, T. E. Furia (Ed.), Chemical Rubber Co. Press Inc., Cleveland, Ohio.
53. SCHADE, A. L. and CAROLINE, L. (1944) *Science*, **100**, 14.
54. MOATS, W. A. (1978) *J. Fd Prot.*, **41**, 919.
55. LAW, B. A. and REITER, B. (1977) *J. Dairy Res.*, **44**, 595.
56. BYERS, R. B. and ARCENEAUX, J. E. L. (1977) 'Microbial transport and utilization of iron', *Microorganisms and minerals*, E. D. Weinberg (Ed.), Marcel Dekker Inc., New York, pp. 215–49.
57. WEINBERG, E. D. (1978) *Microbiol. Rev.*, **42**, 45.
58. WAWSZKIEWICZ, E. J., SCHNEIDER, H. A., STARCHER, B., POLLACK, J. and NEILANDS, J. B. (1971) *Proc. Natl. Acad. Sci., USA*, **68**, 2870.
59. CUMMINS, A. S., DAUN, H., GILBERT, S. G. and HENIG, Y. (1974) US Patent 3 795 749.
60. BENEDICT, R. C., STRANGE, E. D., PALUMBO, S. and SWIFT, C. E. (1975) *J. Agric. Fd Chem.*, **23**, 1208.
61. HUFFMAN, D. L., DAVIS, K. A., MARPLE, D. N. and McGUIRE, J. A. (1975) *J. Fd Sci.*, **40**, 1229.
62. MERMELSTEIN, N. H. (1979) *Fd Technol.*, **33**(7), 32.
63. FIDDLER, W., DOERR, R. C. and WASSERMAN, A. E. (1970) *Agric. Fd Chem.*, **18**, 310.
64. FIDDLER, W., WASSERMAN, A. E. and DOERR, R. C. (1970) *Agric. Fd Chem.*, **18**, 934.

65. HANDFORD, P. M. and GIBBS, B. M. (1964) 'Antibacterial effects of smoke constituents on bacteria isolated from bacon', *Microbial inhibitors in food*, N. Molin (Ed.), Almqvist and Wiksell, Stockholm, pp. 333–46.
66. HOUBEN, J. H. (1974) *Voedingsmiddelentechnologie*, 7(41), 8.
67. HOUBEN, J. H. (1975) *Voedingsmiddelentechnologie*, 8(15), 5.
68. POWERS, J. J. (1964) 'Action of anthocyanin and related compounds on bacterial cells', *Microbial inhibitors in food*, N. Molin (Ed.) Almqvist and Wiksell, Stockholm.
69. JURD, L. K., JR, MIHARA, K. and STANLEY, W. L. (1971) *Appl. Microbiol.*, **21**, 507.
70. SMALE, H., WILSON, R. A. and KEIL, H. L. (1964) *J. Phytopathol.*, **64**, 748.
71. BLOOMFIELD, S. (1974) *J. Appl. Bacteriol.*, **37**, 117.
72. URBAIN, W. M. (1978) *Adv. Fd Res.*, **24**, 155.
73. International Atomic Energy Agency (1978) *Food preservation by irradiation*, (*IAEA Proc. Ser.*), Vol. 1 and 2, Inst. Atom. Sci. Agr., Wageningen, Netherlands, 21–25 Nov, 1977 (Sponsors: Food & Agr. Org./WHO), IAEA, Vienna.
74. WIERBICKI, E., HEILIGMAN, F. and WASSERMAN, A. E. (1976) 'Irradiation as a conceivable way of reducing nitrites and nitrates in cured meats', *Proc. 2nd Int. Symp. Nitrite Meat Pros.*, Zeist, 1976, Pudoc, Wageningen, Netherlands, pp. 75–81.
75. CHICHESTER, D. F. and TANNER, F. W. (1973) 'Antimicrobial food additives'. In: *Handbook of food additives*, 2nd edn, T. E. Furia (Ed.) Chemical Rubber Co. Inc., Cleveland, Ohio, USA.
76. KOHLER, A. R., MILLER, W. H. and WINDLAN, H. M., American Cynamid Co. (1962) US Patent 3 050 401.
77. FURIA, T. E., Geigy Chemical Corp. (1971) US Patent 3 563 770.
78. HOWELL, C. and MARTIN, W. J. (1978) *J. Fd Prot.*, **41**, 44.
79. TWEDT, R. M. and BOUTIN, B. K. (1979) *J. Fd Prot.*, **42**, 161.
80. DAVIES, J. and SMITH, D. I. (1978) *Ann. Rev. Microbiol.*, **32**, 469.
81. WACHSMUTH, K. K., FALKOW, S. and RYDER, R. W. (1976) *Infect. Immun.*, **14**, 403.
82. PINTAURO, N. D. (1974) *Food additives to extend shelf life* (*Food Technology Review No. 17*), Noyes Data Corp., Park Ridge, N. J., USA.
83. IVEY, F. J. and ROBACH, M. C. (1978) *J. Fd Sci.*, **43**, 1782.

INDEX